THROUGH ALL THE CHANGING SCENES OF LIFE:

A METEOROLOGIST'S TALE

Hubert Lamb

TAVERNER PUBLICATIONS
1997

Contents

Dedication	v
List of illustrations	vii
Preface	xi
Acknowledgements	xiii

Chapter

1	Birth and Family Origins	1
2	Childhood	10
3	Schooling	19
4	Cambridge Undergraduate Years	23
5	After the degree examinations	32
6	Towards a Dead End	40
7	Becoming a Meteorologist	53
8	At home in Scotland	57
9	A Visit to Iceland	73
10	The Nations prepare for War	80
11	To Ireland and the new Trans-Atlantic air route	83
12	Interlude	109
13	Going Whaling	114
14	Home from the Sea: Love and Marriage	137
15	My posting to Germany	160
16	And the Switch to Malta	169
17	The chances and changes of life in Research	178
18	Family responsibilities and changing prospects	194
19	First tasks for Climatic Research	203
20	Holidays	235
21	New Directions	249

Epilogue	Language never stands still	253
Index		264

© 1997, M. Lamb

ISBN 1 901470 02 4

Published by
TAVERNER PUBLICATIONS
Taverner House East Harling
Norfolk NR16 2QR UK

To my very dear wife
Moira

List of Illustrations

1. Hubert Lamb c.1975, near the Climatic Research Unit, University of East Anglia.

2. Professor Horace Lamb, c.1919-20, in Manchester, aged about 60.

3. Hubert and Moira Lamb, at home in Holt, Norfolk, in 1994.

4. Catherine, Kirsten and Norman Lamb, the next generation, in 1994.

5. The grandchildren: Adam, Kate, Archie and Tom, and Anna with Catherine and Kirsten, in 1995 and
6. Ned in 1995-96.

7. The author's mother, Lilian Brierley in 1904, as a young woman of 27.

8. The author's father in 1914, aged 36, Prof. Ernest Horace Lamb.

9. The author's father and mother in relaxed mood, c.1938.

10. Sunlit birch tree on a mountainside at Gausdal in Norway, showing 5 to 6 days growth of ice crystals produced by a steady SE breeze in hill fog: as seen on 8 March 1939, after a switch of the wind to W to NW.

11. Fjaerland: end of the road and of the valley and fjord, with the Norwegian ice cap (Jostedalsbre) and glacier tongues in view. (Today the road no longer ends at Fjaerland but continues through a tunnel under the mountains and ice cap.)

12. Chamonix valley in the French Alps, from a painting by Jean Antoine Linck in the 1820s when there were long, extended glacier tongues reaching the valley floor, which have since disappeared thanks to the prolonged glacier retreat.

13/   "The fairy-tale land is here again": pine trees encrusted in ice and
13a   snow, including some in the form of old troll women - Gausdal area, March 1939.

14. *Fata morgana* - mirage deforming views of the north coast of Iceland, near Rifstangi, September 1938.

15. Sailing along the north coast of Iceland, westbound, near Eyafjörður, September-October 1938.

16. Whale-factory ship (Wh/F) *Balaena*, seen in Cape Town harbour, with Table Mountain, November 1946.

17. Strange moonrise in the Antarctic, sketched about 12h. GMT, 15 March 1947, near 64° S 106° E. Wind SW to W, force 5 Bft, temperature -6.8C, weather clear. (About half an hour passed with the pink drapery continually undergoing some changes of form before it became obvious that the object being watched was the moon!)

18. Calm, luminous sea at the edge of the Antarctic pack-ice, under a bright night sky, near 60° S 67° E, temperature -3° C.

19. Smooth swell rumpling the open water surface at the edge of the pack-ice, with wind off the ice and one tabular iceberg: 64 to 65° S near 108°E, wind southerly force 5.

20. "Walrus" amphibian aircraft for whale spotting, catapulted off the factory ship Wh/F *Balaena* over the Southern Ocean, February, 1947.

21. Flensing: deck scene on Wh/F *Balaena* with work beginning on stripping the blubber from the season's biggest (blue) whale carcass, 93 feet (28.4 metres) long by 13 feet (4.0 metres) thick, on 12th January 1947.

22. Tanker m/s *Norvinn*, moored to Wh/F *Balaena* for the transfer, by pumping, of our oil production. Note the inflated whale carcass used as fenders to keep the ships apart.

23. Tabular iceberg in sunshine. Note the caverns worn in the ice at the waterline.

24. Ice wilderness: profusion of bergs and innumerable smaller pieces of ice, after a storm, in 64 to 65° S, near 101° E.

25. Mountainous sea: near 50° S 25 to 30° E, in the zone of the Brave West Winds, April 1947.

26. Rough water in the breadth of the Southern Ocean zone of Brave West Winds, near 50° S 25° E, April 1947.

27. The author with attenders of the Climate and History Conference, in the University of East Anglia, July 1979.

Preface

Having been so lucky as to have had a mostly very happy and fulfilling life, in which moreover the greatest stresses that I had to bear came in the earlier years, I have felt impelled to write this account at least partly as a thank offering for my good fortune. Of course, I encountered difficulties and plenty of worries. But my origins and family background were pregnant with very varied interests, which ultimately produced opportunities, but first I had to learn to live with some quite severe inborn limitations and, at the same time, to cope with excessive expectations on the part of many older family members. Nevertheless, events guided me towards matters, people, and places that I found absorbingly interesting and which led me to a sequence of experiences that I found myself well fitted for.

And I have enjoyed a great bonus on top of all this in that all the latter half of my life and more has been lived in the glow of my dear wife's love and that of our harmonious family.

There are some lessons of this happy experience which are surely worth passing on to others. One is the endless fascination that I have found in just watching the development of weather and climate, and a counterpart of that is the sense of responsibility with which I believe we must view any large-scale human interference with nature. The book records the steps which led to the setting up of the Climatic Research Unit in the University of East Anglia, which is now thriving under its third director, Professor Trevor Davies. My tale also tells how the wider communities of society and the nations sooner or later respond - and can evidently be expected to respond - to a consistently friendly approach. I hope that it may therefore be a modest contribution to social and international understanding.

Acknowledgements

I wish to thank my friend, Professor Trevor Davies, the present head of the Climatic Research Unit, most sincerely for his very kind work in exploring all aspects of the arrangements for publication of this book and for his technical guidance which has been invaluable. Photographic preparation of the illustrations was ably done by Malcolm Howard in Norwich, and I am much indebted once again to Margaret Gibson for the high quality of her preparation of the typescript and to Philip Merrin who scanned the original text for inputting on the computer. Also members of the Climatic Research Unit staff have played a very helpful part in the production of this book, notably Susan Boland and Christine Jeffery.

The view of the Antarctic for the cover was reproduced from a photograph from the personal collection of Julie Hall of the British Antarctic Survey.

Grateful acknowledgement is made to the Controller of Her Majesty's Stationery Office for permission to include in the bookfigs. 10, 13, 13a, 16, 17, 18, 19, 23, 24 and 25, most of which first appeared in the Meteorological Office's Geophysical Memoir No. 94 in 1966. All these figures are Crown Copyright. Permission given by the Associate Editor, Judith C.C. Torres, of the World Meteorological Organization Bulletin, vol. 43 (2) and by Mr Hessam Taba (who took the photograph) to reproduce fig. 3 is hereby acknowledged with thanks. Fig. 27 is reproduced by kind permission of the *Eastern Evening News*, Norwich.

CHAPTER 1. BIRTH AND FAMILY ORIGINS

I was born in Bedford on 22nd September 1913, a pleasant, riverside town whose most famous citizen was John Bunyan, though the authorities of his time put him in the gaol, where he wrote 'The Pilgrim's Progress'. His offence had been to be a non-conformist. My father was working in an engineering firm (W.H. Allen) in Bedford, after serving his apprenticeship first with Mather & Platt of Salford Iron Works and then (about 1902-5) for a time in Hamburg and Hanover, I believe, in Germany. Around the time of my birth he was appointed professor of civil and mechanical engineering at the old East London College, which was later re-named Queen Mary College. Before I was 3 months old we left Bedford, and I never had any further connection with the place, although two uncles and their wives lived in or near the town. My mother's father was a (reportedly irascible) clergyman of the High Church party, the Revd. George Brierley, who had died twenty years earlier. They had lived in Cumberland at Wetherall near Carlisle when she was born, but later moved to Filey on the north-east coast before a further shift to Bedford about 1890. His widow, who also died before I was born, came from a family that owned cotton mills in Lancashire and Cheshire.

My father's father was old Horace Lamb, renowned mathematician, whose pioneering book on hydrodynamics, originally drafted from his lecturing notes in 1879, was reprinted many times and still brought in a small income to his descendants until after 1990, an almost unheard-of longevity for a scientific textbook. His manifestly talented family of three sons and four daughters mostly achieved illustrious careers, a fact of my background which was to worry me in my teenage years, as it became clear how far my parents' and grandparents' , and some of the aunts' and uncles' hopes, had become centred on me as the one heir to the name who seemed headed for any sort of scientific or public career. It was, I suppose, natural that they should think that way. Families did. But how on earth would I ever live up to them and their expectations?

Their attitude was understandable since old Grandpapa Lamb's achievements seemed almost unbelievable given the handicaps surrounding his start in life. He had been born in November 1849 in a mean street of artisans' cottages in the middle of Stockport, outside Manchester. His father was registered as John Lamb, reed-maker for the looms of Lancashire cotton mills, but died of liver disease at the age of 34, in July 1851, before Horace was two years old. And the child's older brother, Harold, had died in infancy in 1848. His mother, born on a Derbyshire farm (Phoside in Hayfield parish) in 1817, worked in the mills, at least until she married again in March 1853. Her new husband, a Mr. Oldham, used to beat her, and in or about 1850 her father, Reuben Rangeley, a Methodist home missionary, was killed by two women whom he was trying to separate when they were fighting in the street. Horace was brought up by an aunt in Stockport when his mother re-married: this lady was kind to him but a very strict Methodist. She allowed him only one toy, a model yacht, which was kept in a drawer and just looked at occasionally.

In spite of this separation from his mother and the disparity in their experiences of life, Horace Lamb was good to his mother in her old age. She was known to the Manchester family as "Granny Oldham", and used to visit them fairly regularly on Sundays. And she for her part was enormously impressed and respectful of her son and his achievements. There was also a step-sister of Horace's "Annie Oldham", who became a splendidly spry little old lady. She visited us occasionally until the early nineteen-thirties, when she was about 82, and used to explore the fringes of north-west London by rides on the top of buses and trams.

Horace's lot first began to ease in the 1860s when the headmaster, Mr. Hamilton, of Stockport Grammar School spotted his unusual ability. Horace, in 1867, when only 17, on his master's advice, declined a classics scholarship which he had just been awarded by Queens College, Cambridge, because his main talent was already thought to be in mathematics. He sat for, and won, a mathematics scholarship to Trinity College the next year. He was second Wrangler in the university examinations in 1872 and a Fellow and Assistant Tutor of Trinity College from 1872 to 1875. He then went out to Adelaide, Australia, on being appointed as the first professor of mathematics in the new university there. He took his newly wedded wife, Elizabeth (Bessie) Foot of Rathfarnham near Dublin, the 30 year old sister-in-law of his schoolmaster, with him. They went out under sail and steam in the usual type of ship of the day and returned the same way ten years later, when he was made professor of mathematics in Manchester University. To my grandmother, and doubtless to Horace Lamb himself, their life in Adelaide was a great adventure, an exciting memory till her dying day.

Although Horace had become a Fellow of the Royal Society before he was 35, in various ways his non-academic achievements were the most surprising aspects of his life for a poor boy of such difficult origin. All accounts agree that he was an unusually attractive lecturer, who made people quickly feel at ease with himself and his subject. From the beginning of his career he seems to have had an easy *savoir faire*. He had a great command of clear, simple English, but also read a lot in French and German and some Homer in the ancient Greek. At the age of 20 to 21 he was a member, under Whymper, of the third party that ever climbed the Matterhorn, only a few years after Whymper's first ascent, in 1867, which ended in a famous accident when, on the descent, the rope parted and half the party fell to their deaths. Maybe it was on the way back from the climb that he found himself in Paris at the time of the Commune, after the Franco-Prussian war. The effect of witnessing this seems to have made Horace Lamb into a life-long Tory. It is not known how he had come to be recruited to the Matterhorn climbing party as a member of such a notable expedition, but it suggests he may have had the benefit of some older person's fatherly interest, though there may have already been a coterie of climbers in Cambridge at that date, some embryonic form of the Cambridge Mountaineering Club. Horace did not continue as a mountaineer, though his children were taken on many hill-walking expeditions in the Lake District and North Wales, the family's main leisure activity.

His eldest daughter, my Aunt Helen, seems to have been the only one of his children who really tried to find out more about Horace Lamb's origins. Maybe the others simply accepted his dismissive contention that family trees were all bunkum anyway and beneath the notice of intelligent people. He had once told Helen "My father came a very long way to settle in Manchester" (or he may have said Stockport). Not satisfied, Helen used the enlarged opportunities of exploration in the motor age to read the names on gravestones in churchyards all over England. But it led to no result. And, some years later, chasing the question the conventional way, with the aid of the national records in Somerset House, I found that John Lamb (who was registered as Horace's father) and his father, William Lamb, before him, were born and christened in churches in various working class districts of Manchester. So there was a mystery, which old grandpa Lamb wished to remain a mystery. Could it be relevant that the Manchester area received a notable influx of refugees, musicians and scientists and others, from the failed Liberal revolutionary risings in Germany and elsewhere on the continent in 1848? The upheaval in their lives, and their anxieties in the months that followed, would surely have been liable to lead to irregular unions. Horace's birth date was close to the time of the greatest likelihood of such contacts. It is relevant to add that his stolid good health throughout his life made a sharp contrast to what we know of John Lamb, reportedly his father, whose other son died while still a baby.[1] Horace's mother seems to have been a healthy, spirited girl from the countryside, who was in trouble with her second husband soon afterwards. Horace, moreover, was physically of a Germanic type, sandy haired, healthy-looking, with a fresh complexion. He jokingly told his dark brown haired and dark complexioned Irish wife, Bessie Foot, before the family: "I am the Saxon, and you are the Celt".

It is appropriate to begin this account with the background of the family and its great man in this way. We see them emerging from social mists in the middle decades of the nineteenth century and outer darkness before that. There were tales of the prosperous set-up of the Foots and their tobacco business near College Green, Dublin, and of dinner-party guests arriving by coach up the torch-lit drive to their house, Holly Park, on the lower slope of the Wicklow Hills, near Rathfarnham. But by the eighties old Horace Lamb was revered by them all as the founder and father of the family. I only knew him after his retirement. He had become a kindly, though rather silent man. He certainly appeared genial to me, and we occasionally played board games together - backgammon, Halma and Reversi, I remember. He had moved through life charming most of those he met, not least the ladies. He was described in the obituaries as having had "a sunny character". And everybody was struck by his lucid brain. But there was a hard side to his nature too, as was not uncommon among fathers in his time. Two of his seven children were in the end cut out of his Will: Peggy for becoming a Catholic nun, when she was in shock on losing her fiancé in the Great War about 1916; and Henry, the artist who was finally elected to the Royal Academy, for leading what his father regarded as an

---

[1] Despite these suspicions which, considering particularly Horace Lamb's own dismissal of family trees and his children's inquiries about family history, seem to offer the likeliest explanation of the known facts, a different explanation of the source of his genius was put forward around the time of Horace's death. John Lamb was described as a prosperous reed maker for the looms of the cotton mills, who received advancement because of his invention of an improvement to the reeds. This is hard to reconcile with the other recorded facts. He had flitted from one address to another in Stockport, first in John Street, then Steel Street, and finally Chestergate, all poor working-class streets at the time. John Street was one of the meanest streets in Stockport, lived in by the basic grade of mill workers. He was married in a back street church in Manchester and his widow had to work in the factories.

Birth and Family Origins

abandoned life in the "Bloomsbury Set" of the same period, after giving up his medical course at Manchester University against his father's wishes. These events caused repercussions all through the family for the rest of the lives of my grandfather's children. And in due course it was my fate also to be left out of my father's Will.

Helen, his eldest, after teaching at Bedales and being early widowed, became for several decades head of Peile Hall in Newnham College, Cambridge. My father, the eldest son, as already stated, was a professor of engineering in London University. Walter was Secretary of the Royal Academy of Arts from 1913 until 1950 or after. Henry, the artist, who had finally completed his medical course in order to serve as a doctor at the front in the Balkans and later Palestine in the 1914-18 war, was the youngest son. Lettice, the second girl, was a nurse, widowed in middle life when her doctor husband working in cancer research died of the disease. After Peggy, the nun, came finally Dorothy, the youngest, born in 1887. She had first worked in archaeology and became a lecturer in England and the United States, before marrying Reeve Brooke, brother of Rupert Brooke, the poet. She found herself wife of the vice-chairman of the Electricity Commission and later senior civil servant in the Ministry of Transport. After his death she married his friend, Sir Walter Nicholson, head of the Air Ministry. These people over-awed me completely in my youth.

Some other memories from my early years of family and friends are needed to set the scene in which I grew up. My mother's family were a very different lot from the Lambs. Her family were cotton mill-owners in Lancashire and Cheshire. And it was that background that enabled the Bedford family to survive on a modest income after the clergyman father died of his heart attack in 1893 and even to send my mother and a sister to the High School. Her elder sister, Edie, was a rather gypsy-like, good-hearted, generous character whose generosity exceeded her means - she and her RAF dentist husband and her two children all also smoked "like chimneys" - so that my thriftier parents were for ever paying her bills. But it was her brother Reggie and his wife, Auntie Louey, a childless couple, who were my best friends in our family. During the war they lived in London (Kensington) and my mother and I used to make long journeys on the 28 bus to spend a day with them. It happened several times that the temporary wartime bus drivers were uncertain of the route, and this cocky 4 to 5 year old "saved their bacon" by telling them which was the right turning to take. My uncle and aunt were not well off, but had a small "Rover 8" car in the nineteen-twenties and used to come and stay with us for some days twice a year. And each time I got half-a-crown extra pocket money from my uncle, a highly valued gesture which helped to endear them to me. He was county architect in Hereford, and from the age of 12 upwards I was sent by train to visit them on my own, my first visits away from home. Those visits were great fun. I was driven around the countryside in the "dickey-seat" that opened up in the back of their little car, when my uncle was visiting schools on business. The trips were liable to be extended occasionally for a picnic on the Malvern Hills or into Wales for a picnic on the Black Mountains and exploration of the ruins of Llanthony Abbey. My mother had another brother, Uncle Bertie, who married but also remained childless and spent all his life in or near Bedford. And my mother suffered a severe blow in 1932 when both brothers died of heart attacks in the same year. That was also the year when her lifelong closest fried, Elsie Barlow-Smith, from her school days in Bedford onwards, but who happened to live near us again later, died of cancer. She had married a Captain Huddleston of the Royal Indian Navy. And so it came about that her son Trevor Huddleston, who later became Bishop Trevor Huddleston of South African fame, was the closest friend of my childhood. Both Trevor and I were brought up as regular churchgoers and went to Sunday school teaching in the Anglo-Catholic tradition. Our attitudes began to diverge, however, when we were just 10 to 11 years old, dressing up in his playroom. He liked to play "church", robed in make-believe priestly vestments and acting ceremonies, but this troubled me and set me thinking "what is this all about?". Nevertheless, his attitudes and mine on the social demands of Christianity, when we grew older, have been very much alike.

At the end of this stage of my schooling, when we were nearing 14, Trevor's father presented the school with a strange gift, a metal cup like sports cup, a "VOX POPULI" to be won each year by the boy who was voted by the others as the most popular boy in the school. To our surprise, since neither Trevor nor I were much interested in the usual sports, Trevor won it that first year and I was the runner-up.

Another school friend of mine at that same school when I was between 8 and 14 years old, was Olaf Richardson, the adopted son of Lewis Fry Richardson, the distinguished Quaker meteorologist, who resigned from the Meteorological Office in 1919 when it became part of the Air Ministry. In spite of that, he went on to develop the first computed weather forecast. It was before the days of the availability of computers, so Richardson was given the chance to employ forty-eight meteorological assistants to do the calculations from a network of observations of the state and movement of the atmosphere at the starting time. It took three weeks for them to compute "tomorrow's forecast". The answer was not a success either, but it has been recognised as the first, pioneer attempt which paved the way for what is now a daily routine that has also greatly improved the success of the forecasts.

Birth and Family Origins

This was the beginning for me of a long acquaintance with the Richardsons, with Lewis and his work and character, his wife Dorothy who often gave me tea after school with the family, and with the sister Hilda who favoured the same Anglo-Catholic church as my parents. Lewis Richardson was a birthright Quaker whose antecedents went back to the earliest days of the Society of Friends (Quakers) in the seventeenth century. Early on in his career he had seemed to have amorous intentions towards my father's sister, Dorothy. But when I visited the Richardsons many years later in Paisley, where he was the principal of the Technical College, I became aware, much more than before, of all the self-denials of artistic things and pleasures, which left a very Spartan self-sacrificing interpretation of the demands of Christianity that frightened me off joining the Society of Friends for many years. I was, however, clearly drawn to a Quaker understanding of life, and I did join in the end, though I have never felt called to the comprehensive sacrifices of the colour of life that Lewis practised but evidently with some potential dangers. I myself still enjoy a fine church service, grand or devout, with carefully chosen, but not too much music - such as is often to be had in Church of Scotland town churches, the Lutheran churches of Scandinavia and in some Church of England churches. I feel really indebted to my parents and those among my friends, notably the Huddlestons and the Richardsons, who, by their example, showed me the prime place that religion should have in our lives and how it enriches one's life if it is accorded that position.

There were others among the long-standing friends of the Lambs, who brought stimulating and educative contacts with other countries, especially Scandinavia. Chief among these were the Fluxes. Sir Alfred Flux, born in 1867, thus older than my father and mother, was an internationally respected economist, Senior Wrangler in 1887, a Fellow of St. Johns College, Cambridge, and much later, for a time, President of the Royal Statistics Society. He had held professorships in Manchester and Montreal before becoming a leading statistician with the Board of Trade in London, and he worked for many years for the League of Nations in Geneva. But it was his Danish wife, Emilie Hansen, known to me in my babyhood as "Auntie Fux", who was the darling of the whole Lamb family from her days in Manchester in the 1890s. They later lived in London until 1932. On a visit to the Fluxes in 1937, three years after Horace Lamb's death, Lady Flux remarked to me "What a strange family the Lambs have turned into. When they were only 14 or 15 years old in Manchester, there was always so much fun in that household.[2] And now, how different they are.....they have all grown apart. The brothers never see each other, and the sisters too. Lettice was cruelly twitted and teased by Helen and Dorothy." Indeed, Lady Flux told me she had witnessed this going on even in our house, when visiting us in the Hampstead Garden Suburb. "And Lily (my mother) was quite unable to defend poor Lettice. That is why Lettice prefers to stay in Manchester. There is far too much emphasis on brains in the Lamb family." This was perhaps the earliest direct comment to me on something I have too often since observed myself; how ambition, jealousy and scorn can ruin friendships and family life. And it came from someone who had had a close, even fond, relationship to three generations of our family.

On that visit to Copenhagen in 1937 I was staying at the youth hostel, in a huge dormitory with 25 to 30 beds, upstairs in the Young Men's Christian Association (KFUK) and witnessed an incident that revealed the pressures that young Germans were under at that time. It turned out that nearly all my fellow-hostellers were Germans from Hamburg. A couple of them got up early, around 6.15 a.m., stomped down the long room, their right arms stiffly stretched out forward in the Nazi salute and barked out

---

[2] If the rumours I have heard have any solid foundation, some of that uproarious fun perpetrated by the teenage Lamb boys in the eighteen-nineties went far beyond the bounds of any responsible behaviour. This must be always liable to happen with high-spirited boys of that age before they are fully aware of the dangers and losses to the innocent public that they might cause. Some of the stories are therefore, to this day (and maybe particularly in those days), best not retold here. But one, probably the most harmless, of their pranks (or what some of them were involved in) is worth re-telling. The identities of those who planned and executed it have never been divulged. This was the pretended visit of the "Sultan of Zanzibar" to Cambridge. Three or four friends from Cambridge and Oxford got together and, having sent a telegram to the Mayor of Cambridge a while beforehand, duly descended upon the university town in appropriate clothing and with proper solemnity. They were met at the station and officially welcomed by the Mayor and dignitaries of the town and university. A suitable programme had been arranged for the distinguished visitors with banquets and speeches, before the party left by rail as they had come and disappeared into oblivion and anonymity. Such grand-scale tomfoolery would hardly be possible in the days of radio with the whole modern communications networks and the ubiquitous reporters of today.

"*Heil Hitler*". There was no response. As soon as they had gone, 22 other young Germans set about apologising to everyone else for their compatriots' behaviour. They said they had all belonged to the Scouts before they were lately forced to transfer to the Hitler Youth. A little later I had a look at the hostel Visitors' Book: there were two signatures with "*Reichsdeutscher*" (imperial Germans) in the nationality column and 22 others who had entered themselves as "*Hamborgere*" (Hamburgers) in the same column.

The Fluxes asked me what young people of that time in England were thinking about the threatening state of the world. I said "I am shocked. Mostly they don't seem to think about it." Lady Flux said "It is the same here in Denmark. But I believe it is better in Norway and Scotland. There is less ridicule there for serious-minded people." I was disappointed to discover how aggressively British 70-year old Sir Alfred Flux still was after four or five years of working at the League of Nations in Geneva. His wife thereupon told me that he just regarded war as a recurrent phenomenon, as if one could never hope to do anything about it. But she was above all impressed by how the First World War had made people uncivilized and brutalized their thinking and behaviour. "How brutal the talk has become in all countries". I completely agreed. And by the 1990s how much more the brutalization has proceeded. Already in the 1930s, some of my young friends in Norway were talking of "*tidens brutalisering, tidens tilbakegang*" - of civilization in retreat. And the wars and teaching the young to kill were recognized as the main reason.

CHAPTER 2. MY CHILDHOOD

Such were the blood relationships and inherited influences behind my upbringing.

My earliest memories are of my mother's anxieties as she coped alone with me as a tiny child while my father was away at the war, and she worried because for the first two years I could hardly keep my food down (possibly an indirect result of my rupture (hernia) operation at the age of 4 months). I have a clear pictorial memory of my father's return to the household in 1915 after the Gallipoli campaign, and just where I and my mother separately stood in the hallway of our house, waiting till the front door opened and this unknown man came into the scene to join our little household. For the first year or two his arrival, however, made no real change to my circumstances that I can remember. He was busy and much of the time away at the Naval School of Mines in Portsmouth, where for the last year of the war we went to join him.

My memories of the "Great War" consist mainly of a few aeroplanes, always bi-planes, in the sky, the markings occasionally German, and several bombing raids, when the bombs dropped fell within half a mile of our house, luckily always in open fields beyond the oak wood that was next to us, but quite near the last road of houses at the edge of the suburb. I was taken on some afternoon walks beside my push-chair to see the holes where the bombs had fallen. During the raids we had crouched together under the double-leafed dining room table on the ground floor. One night, after I was in bed, my father wakened me to see (from a top floor balcony) the horrific sight of a Zeppelin coming down in flames near Potters Bar, five or six miles north of where we were. I was mainly conscious that we were witnessing a tragedy produced by man-made failures of governance.

My other memories of that time are all pleasant. I count myself lucky to have been born in time to experience the last years of the old horse-drawn age, when life went to a quieter pace. We were only five miles from Charing Cross but yet at the edge of London, with a narrow strip of open country still leading away to the north-west. I was lucky, too, with the outlook from the windows of my upper floor nursery, looking out through young acacia trees to the water-bound road edged with wooden kerbs. At the top of the hill was the nicely laid-out central square of the Hampstead Garden Suburb with the two big, recently built brick churches - the Anglican one with a great steeple and the United Free Church with a central dome. Away at one corner of the square, under some oak trees at the edge of the wood, was the Friends (Quaker) Meeting House, a neat single storey building with a sloping tiled roof. All this central part of the suburb had been architect-designed by

My Childhood

Lutyens with the then rising, post-Victorian respect (which his work was building up) for English rural cottage traditions. Regular sights were the lamp-lighter coming at dusk on his bicycle; with his long pole, to light the gas in the lamp-posts on either side of the street; and, doubtless after the war, the muffin man on Sunday afternoons with his open tray on his head and his bell in his hand; and the milk cart with its great deep can like a brightly washed and polished saucepan and its scoop-ladle. The dustmen, the coal merchants, and Garter Paterson (the luggage and parcels delivery), had very much bigger carts or horse-drawn lorries. Bread was delivered in quite high, narrower, covered carts of intermediate size. Other objects in the scene were a massive, cement-like drinking trough at the cross-roads for the horses and the similarly built receptacles with hinged metal lids for grit to help the horses in snowy or icy weather.

The 1916-17 winter provided weeks of frost and snow, often with fog and rime on the trees. I remember peering out at these new sights through nursery windows decorated with "frost flowers" formed by the ice condensing on the insides of the panes.

The Hampstead Garden Suburb community in those early days was a pleasant and, I think, healthy mix of most social classes. The next road was one of smaller houses and lined with crab-apple trees, each house also having its garden. Dame Henrietta (Octavia) Barnett, wife of Canon Barnett of Toynbee Hall, Whitechapel in east London, creator and founder of the garden suburb, was living when I was a child on South Square at the top of the hill. It had been her initiative that had collected funds and bought the land for the Hampstead Heath Extension, to preserve it from a "builder's rash" that would have ruined Hampstead, when the Underground was extended north of the high ground, and the heath, to Golders Green. The success of this owed everything to her social vision and prompt reaction to a chance remark which she and her husband heard on board a ship bound for Russia in 1896. Indoors, too, most of the furnishings in our house and neighbours' houses were of simple straightforward designs which no doubt helped train my eye and, maybe, my thinking on other things besides.

In our area, co-education at least in the starting schools and in some cases even up to the older age-groups was the norm. It was a peaceful community surrounding us and a charming place to grow up, which doubtless encouraged visions of a harmonious society. Yet it was also an unusually politically concerned place. Among those living nearby were more than a few Quakers and several leading Liberal and Labour politicians, as well as people with other kinds of involvement in 1920s national and international life. Among the most memorable neighbours for me were the Tudor-Pole family, Conservatives, who lived for some years directly across

My Childhood

the road before moving to a house-boat on the Thames. The father was rumoured to be something in the Secret Service. The mother was a very striking, generous lady with a great mop of blond hair, who had interesting friends in the Russian émigré community and favoured both a very high-church Anglican church in Camden Town and contacts with the Russian Orthodox Church. They espoused some unconventional "progressive" educational establishments while inclined towards, and later active in, right-wing politics. As somebody in my youth neatly put it: "Consistency is the hobgoblin of little minds".

It was when my mother and I moved to Portsmouth (Southsea) to be near my father in about the last year of the war that things seemed to get more difficult for me. It must be impossible now to know how far my father's severity towards me was due to tendencies in his own character, and his tendency to follow blindly the precepts of an earlier age, or how far he had been psychologically damaged by the war. There seems no doubt that his pride had been hurt, and possibly his self-confidence wounded, already many years earlier, when the second son, Walter, was treated by old Horace Lamb as the brightest child and succeeded in going to Trinity, his own old college, and becoming a Fellow of the college from 1907-13. My father, who was considered a rather colourless character by his brothers, had to be content with Owens College, Manchester. Walter had even been invited to discuss with his father whether to do classics or maths, and was firmly supported when he chose classics. This only seemed to me to show how much more enlightened my grandfather was than my father, who gave me no choice and never discussed such matters with me. But I suppose it had been a bitter blow to my papa's self-esteem. It may also be that old Grandpa Lamb had a secret love of languages and the classics in particular. Until his last years he kept his copy of Homer, in the ancient Greek, at his bedside. And this may have played a part in his readiness to support Walter in choosing to do a classics course.

Certainly, my father had volunteered to join the forces in 1914 with more than the average enthusiasm that prevailed even at that time. And he was soon prominent for his hatred of the Germans, whom he always referred to as "the Hun". But he did not suffer from a gas attack as his brother, Henry, did. He was a great stickler for instant obedience and considered me a particularly disobedient child. This resulted in frequent "thrashings" or "tannings", as he called them, and many repetitions of the dogma: "Spare the rod and spoil the child". One Saturday evening, soon after we had joined him in Portsmouth when I was just 5 years old, about tea-time, he took me out walking ten or fifteen minutes to the shops to choose and buy a cane. I cannot now remember whether I had first committed some misdemeanour

My Childhood

that demanded quick retribution or whether this was just to equip himself to deal with matters whenever they arose. Certainly, the "spare the rod" dogma was several times recited as we walked and beatings did follow at fairly frequent intervals until I was over 16 years old. The results of all this were untold harm to the relationship between us, and even the beatings that were then a routine part of "public school" discipline made very little impression on me. I had acquired an immunity and become sceptical of most authorities. And it remains a mystery to me how a man of his intelligence could not see the actual effects that the treatment was having.

But he was blind to such practicalities. He prided himself on belonging to the "old school of thought". Psychology was to him a dirty word. In his later years I was sorry for him. The authoritarian aspects of the Catholic organization of religion appealed to him and were used to sanction his own authority, permitting no questioning or discussion. And so in my later teens, and especially when I was away from home at university in Cambridge, I took the opportunity to look at other forms of worship and teaching, Roman Catholic, Presbyterian and Quaker, chiefly. And when I was about 20, and had established links with the Presbyterian (now United Reformed) Church there, my parents dispatched me during one vacation to spend several days with a priest, a Father Briscoe, whom they had known well in their Bedford days before 1914. So off I went to the pretty Somerset village of Bagborough among the Quantock Hills to be corrected and re-instructed in "the true faith". But I only argued with him the whole time and it made matters worse.

Whatever else the stern upbringing achieved or failed to achieve, it taught me the need to be self-reliant and probably committed me to being something of a "loner" because so few others of my generation - and none of my acquaintance - had any similar experience. I am sad to say, though it comforted me at the time, that my father was branded accordingly by sympathetic relatives as "a bit of a tartar".

When I was 8, I went with Trevor Huddleston to a private, fee-paying preparatory school, called Tenterden Hall, two miles away, in Hendon. At first, I went as a day boy, the journey to school being in two stages, first by tram north to Finchley, about a mile, and then a walk over the hill beside a show-piece Express Dairy farm with grazing cows, and then along a narrow cinder path between iron railings, and bramble bushes, that ended beside the school. Nobody worried about bad characters on the prowl in those days, although schoolboy legend had it that an old tramp, who was occasionally seen near the path, made his living by fashioning pips for raspberry jam from bits of wood lying about!

My Childhood

I was very fond of my mother, but I realise that once started on serious schooling I became probably unbearably uppish towards her. She, I am sure, did not have the physical strength or mental agility to manage a wilful child. When my father was away at the war, she had employed a succession of young, and reportedly flighty, girls as nannies to help her. None of them lasted very long or made any impression on my memory. But from the beginning of our time in Portsmouth to be near my father, when I was 5, until I was 8 years old, I had a very different nanny, a kind, auburn-haired girl, Florence Stevens, then in her twenties. She took over from my well-worn stuffed dog, "Jessamie", the responsibility of listening to all my tales of woe when suffering from my father's stern discipline. And she managed the delicate task with consummate skill. She used to soothe me at bedtime by singing a hymn ("Our blessed Redeemer, ere He breathed his tender, last farewell...." was a favourite) and reading me a story. She remained a friend of the whole family even after she left and in fact until her death in Hastings in old age in the late 1960s. Before coming to us, this kind, devoutly Christian, lady had been employed by Lady Scott as nannie to Peter Scott when his father went on his last expedition to the Antarctic. And we heard alarming accounts of the unquestioning principles governing Peter Scott's childhood: how he was always clad in minimal clothing, always had a snively nose and suffered endlessly from colds.

This indirect link with polar exploration undoubtedly stimulated some interest on my part. Military leaders never figured among my youthful heroes. Polar explorers, and to a lesser extent Everest climbers, did. It was quite a common interest during those years after the war, as it had been also after the defeat of Napoleon a century earlier, and many books on the theme were appearing. But I was always more impressed by Shackleton, who brought all his men back alive, and by Amundsen and Nansen, than I was by Scott, whose planning seemed less sure. Fridtjof Nansen wrote a book about "The First Crossing of Greenland" that was one of the first books I bought for myself. It was his account of his own expedition in 1888, richly illustrated with his own drawings, some of them copied from photographs. Despite his attractive qualities Nansen, too, was something of a disciplinary extremist: no tea or coffee was allowed on the expedition, possibly because he did not wish to risk having his men's real condition masked by their effects.

When Florence Stevens left, I was sent, aged just 8, to board at Tenterden Hall School, where my friend Trevor Huddleston was already a boarder. But after two terms I made such a fuss about boarding away from home, and having some dim notion that my father who had been a day-boy at Manchester Grammar School did not really approve of boarding schools,

My Childhood

that I was allowed to revert to going each day to school. That I now did by bike and on cold mornings in winter I filled the handle-bars with hot water, held in by corks at each end.

It was after biking home in the afternoons that I sometimes went to tea with the Richardsons, who lived beside the direct route. There I was brought once again in touch with polar exploration because of Lewis Richardson's admiration for Nansen - their dog was called after him - and his general interest in Norway.

My father plainly suffered a long succession of disappointments in me, which he found unacceptable. I had none of the natural abilities which he regarded as essential. Most conspicuously, I simply could not emulate his carpentry and joinery skills. I seemed unable to saw a piece of wood straight, and I never managed to plane wood and produce a smooth, level surface. He himself loved working with wood and was very good at it. He produced some fine items of furniture, including a gramophone case as well as cupboards and shoe-racks and a table with "Chippendale" ornamental features to stand the gramophone on. He wanted me to be his carpentry assistant standing by, sometimes for several hours, to pass him whatever tool was required and sweep up the sawdust and shavings. This inevitably became boring to me when there was so little action needed from me which I could actually do. Moreover, I was never any good at standing for long periods. In his joinery and in engineering jobs, maintaining his car, adjusting its brakes and lights and decarbonizing the engine, he found great relaxation. His strong ambition for me was that I should acquire all the same skills that he had, which he had learnt firstly as a schoolboy in Manchester and later in Germany. In the end, I naturally drifted away from these sessions to my own ploys and his dismay, though to some extent he was fascinated by the simple, unmechanical toys that I made for myself; matchbox or cardboard cut-out buses and trams that could be pushed along vast street maps drawn on drawer-lining paper or tramlines made as grooves in layered cardboard. My efforts were always aimed at a satisfying artistic impression, his at mechanical working.

He and I did, however, share a strange but equal failure at cricket, when he dutifully took me out to introduce me to the game. That part of the Englishman's religion was clearly one thing that neither he nor I were cut out for.

Piano playing was another disappointment. He was good at it and a good Bechstein upright piano stood in our drawing room. But my fingers seemed unable to obey my sight reading. I started piano lessons at school on my eighth birthday and was billed to play a simple Bach minuet in the school's Christmas concert at the end of the first term. When the day came,

My Childhood

my nerve broke, and I refused to play. The result was that I was never allowed another piano lesson, although I spent many hours for many a year sitting at the piano practising and trying to play reasonably simple arrangements of pieces that appealed to me (and still do) by Schubert and Handel, Mozart, Bach, Beethoven and Brahms, making some progress but always liable to accidental wrong notes. It is interesting to me to notice the constancy and persistence of my musical choices, how I still delight most in the same composers and some of the same pieces as when I was a boy of under 10 years old - save that I would now add some Haydn, some Grieg, some Schumann, Mendelssohn and Sibelius, Nielsen Saint-Sans, Berlioz and Bruch, as well as folk songs and folk music from all over Europe. Association with the preceding generations of my family has left me a music lover, but here again put in the shade by the talents of my father and, especially, his brothers. One of the German refugees in Manchester during their childhood and youth was Carl Halle, later Sir Charles Hallé, founder of the Hallé Orchestra.[3] Henry and Walter used to spend many an hour with him and the orchestra at rehearsals. Both the boys were uncommonly able musically, but Henry especially turned into a very gifted pianist himself, capable of good concert performances and did, in fact, give some. It was always a pleasure to listen to his playing.

There were, thank goodness, some happier phases in my relationship with my father, never more so than when as a child from 6 to 9 years old, he used to read to me the old Norwegian country folk tales and legends of Åsbjørnsen and Moe about the trolls and the great hills and forests of southern and central Norway. They had appeared in an English edition in 1880 - a marvellous edition richly illustrated by woodcut drawings. The thrills and the special humour of those tales excite me to this day. All the Lamb boys had copies of this precious volume. I am also indebted to my

---

[3] The Dictionary of National Biography records that Carl Halle was born at Hagen in Westphalia in 1819 where his father, Friedrich Halle was organist and music director at the principal church. He showed a remarkable gift for pianoforte playing even as a child and gave a performance in public when only 4. He studied under Weber in Darmstadt at the age of 16, and in 1836 went on to Paris, where he mixed in the best musical circles with Chopin, Liszt, Berlioz and others. It was the revolutions in France and elsewhere in 1848 that drove him to England. He chose to settle in Manchester because of the influential colony of music-loving Germans there and it became his home for the rest of his life.

father for introducing me to much fine music through his gramophone record collection. And, even on questions of world affairs he and I did occasionally see eye to eye. For instance, he was very hopeful of the calmer period induced by Stresemann's social democratic government in Germany in the middle nineteen-twenties, but he was quicker than I to spot the turn for the worse in Germany after that. I was alarmed to the point of despair over how the French revenge policy of demanding continued payment by Germany of war reparations after the disastrous inflation in 1923 was having the predictable effects of snuffing out the growth of liberal and social democratic tendencies in Germany.

Although my father was born in Adelaide, South Australia in 1878 and spent the first seven years of his life there, and my mother was born in Penrith in Cumberland and spent the first two-three years of her life in Wetherall, a village beside the River Eden, forebears of both families had strong links with one parish, Hayfield, in Derbyshire, a few miles south-east of Manchester. Some were buried beside each other in the churchyard at Hayfield. But it was my mother's ancestors who were the grand folk, the squires whose descent could probably be traced back to Viking stock in the Norman earls of Chester, whereas my Lamb grandfather's mother's people were ordinary working people of the parish. Both families were in the eighteenth and nineteenth centuries perhaps unusually mobile for the times. And the Foots, my grandmama's parents' family, had arrived in Ireland with William of Orange and taken part in the battle of the Boyne in 1689. There was not much really Irish blood in their ancestry. When the Dublin tobacco business declined later in the nineteenth century, reputedly because the sons who inherited the firm overlooked the change of fashion away from snuff, most of the Foots emigrated to Canada and Australia.

The lady, Rose Brooks of Chicago, whom Walter Lamb married, at the age of about 45, in 1927 had an even more mobile ancestry. Her Brooks forebears had had a firm in Birmingham about the beginning of the nineteenth century, which had succeeded in landing a contract from the Tsar of Russia to supply all the ornamental brasses for the harnesses of his cavalry. The eldest son of the firm went to live in Russia to oversee the business and, while there, married a princess of German ancestry and acquired a considerable estate. But when the first widespread troubles with the workers on their lands broke out in 1861, they saw the "red light" and, despite the liberation of the serfs, decided it was wise to sell up their possessions and emigrate to the United States. For a time, therefore, they were well endowed, but later feckless members of the family squandered the wealth despite the efforts of their womenfolk to keep it secure. My aunt was

My Childhood

training as a commercial artist in France when she met and married my uncle.

CHAPTER 3. SCHOOLING

I suppose my schooling should over-all be regarded as a success, even though I did not follow the same plan for my children.

I remember the repeated discussions my father and Trevor Huddleston's father had on the matter on Sunday mornings, walking home from church. The Huddleston's chose Lancing College, largely, I believe, because of its Anglo-Catholic tradition. My father was most impressed with Oundle under its most famous headmaster, Sanderson, who substantially re-cast its development in his time between the 1890s and 1922 with what was then a new emphasis on science and engineering, along with practical experience of a week at a time in wood or metal workshops, as well as modern languages. All these were ideas that won my approval, even if I was not handy in the workshops, and even in my teens I wilted after hours standing beside a lathe. Clearly nothing but the "public (boarding) schools" came into their consideration.

My father had spent a lot of time and effort training me in such things as table manners. Looking back, I have been surprised to recognise how much of conventional etiquette is based on class prejudice. A give-away sign of this was papa's oft repeated ban on spreading the butter over a whole piece of toast, explained by the remark: "You don't want to look like a brick-layer". The same principle seemed to underlie the guidance that you should never eat every last scrap on your plate: that looked greedy (but it also looked as if one could not afford to leave anything uneaten).

There were difficulties, particularly for me, in being away from home at boarding school. I had to cope with two different authorities over my life. And the two disciplines and their underlying principles did not always agree. Moreover, I had been inured to so many beatings at home that the cane had been thoroughly devalued for me. This obviously led to indiscipline and devilment in such matters as drinking an occasional glass of cider in a pub at the edge of the town and slyly smoking an occasional cigarette of the kind then supposedly good for hay fever. These were regarded as serious offences at that time.

But what was really the most serious crisis in my school career arose in quite another way. About the time of my fifteenth birthday, after success in the Schools Certificate examination at the end of my first year at Oundle, I returned from the summer holiday to be told, without warning, consultation or explanation, that I had been switched to the science side and would give up altogether several subjects that interested me: notably history, Greek and Latin. I got on well enough with the mathematics and the physics and chemistry classes that I now started, but I was not ready to accept the

Schooling

specialization involved in the change. My resentment led me into fresh crosses with authority, though I welcomed the "Science German" that I then began: in fact, it took the form of just one lesson a week reading Grimm's Fairy Tales. Another reaction was that I began taking more decisions for myself as to what interested me. In particular, I took to spending hours, when I had time, mostly on Sunday evenings, browsing in the *Encyclopaedia Britannica* in the house library.

One of my worst memories is of the iron rule that forced me all through my first summer at Oundle to spend all afternoon on three days a week, and two hours on the other days, on the cricket field and in cricket nets at which I was useless. The fact that it also aggravated my hay fever was a minor additional irritant. After that first year, however, one was allowed to reject cricket in favour of rowing, and to the river I went. After that my memories of the school summers are of sunny afternoons on the river and gliding by the gentle scenery of osier scrub and tall trees and the gliding motion of the boat, and of having to watch the odd clegg or horse fly settle on one's forearm and not being able to do anything about it. We did not do brilliantly in the races, but there was plenty of enjoyment.

Another bad memory, but in this case just a ridiculous one, was the compulsory soldiering in the Officers' Training Corps. Its field days were memorable only for the opportunities for burlesque and making a nonsense of the whole thing. In this, too, there was a recognized let-out from the second year onwards. I joined the Scouts and became Assistant Scoutmaster to the local scout troop in Aldwinkle, a neighbouring Northamptonshire village. To get me there the Scoutmaster, Dudley Heesom, of Oundle used to drive me the seven miles in his open-topped "Baby Austin" car, the original Austin 7 horsepower. Those runs were fun, especially one autumn evening when we could have caught a pheasant by the legs as it flew low over the car: we desisted because the ensuing struggle with a strong pheasant while the car was being driven could have been dangerous. The country village scout troop were easily managed. I got quite different impressions a year or two later when acting as Assistant Scoutmaster to a much quicker-witted troop of London boys in Islington.

I had tremendous fun out of scouting, especially out of the experience of scout camps, and greatly appreciated its training in simple practical pursuits and co-operation with others. Cooking around the camp fire always fostered a good appetite. One of the most widely enjoyed activities was tracking games. Even the fifteen minutes or more which I wasted one warm summer evening in one of the Yorkshire dales, tracking a small steady light which I saw first when it was several hundreds of yards away, and when I came near without having disturbed it I found it was not

Schooling

one of the opposing team waiting a chance to cross our path but just a glow-worm. Some of the walks across the hills were splendid training in map reading, especially when one was in the mist. But one misty day in August on the top plateau, which was nearly flat - apart from the divided "peat hags" - I had an anxious passage. One of the Oundle boys in my patrol, about the same age as myself and bigger than I, despite our compass, disagreed with the way I was going and wanted to split the party in the thick mist or at least split off himself and go in a different direction. He, and any who went with him, might have spent many hours wandering in the mist. So, up there in the gloomy fog, I had to insist on the dangers of what he was proposing and that the party must stay together. Luckily it was not more than about half an hour before we came out in the clear and were all agreed on our bearings.

One unforgettable interlude in my Oundle career was the long frost in February-March 1929. The Headmaster, K. Fisher, was a keen skater, and he made every day a half holiday while the frost lasted and the Tuesdays, Thursdays and Saturdays that would normally have been half-holidays became whole holidays throughout the four-week span of the frost. We certainly got plenty of exercise. The days began and ended with a walk to and from the ice, either 1½ miles to a very pretty ornamental lake, with woods and reeds at the edge, beside the Benefield road to the west, or 1¼ miles to the flooded meadows beside the river east of the town. Most days were cloudless. The thin covering of snow, that had come on 10-11th February as the frost began, mostly disappeared by evaporation. And the ice on the meadows was entirely safe because the floods soon froze to the bottom. One or two masters skated miles down the river, but that was rightly forbidden to the boys, most of whom were beginners. The fun came to an end with one brilliant Sunday in mid-March when the warm sun had melted the ice all round the edge of the lake, and one had to jump from the grass on to the still thick plate of ice that covered most of the water. The air temperature measured in the shade that day rose to $70° F$ ($21° C$), but the ice had been maintained by sharp frosts every night.

I learnt one more important thing about myself during my Oundle years; that although I won the Headmaster's Maths and Science prize in my final year, my ability at mathematics was going to be limited by my inadequate short-term memory. This meant that in working with complicated equations, when involved with long strings of terms, I could not remember what each term in the equation meant without a terrible loss of time spent in checking and reminding myself.

So in my last winter at Oundle I went by train to Cambridge and, as usual somewhat overawed, staying with my grandfather in the security of his

Schooling

fine residence in Selwyn Gardens, I sat the scholarship examination for Trinity College and ended up with an "exhibition". This success led to grandpapa Lamb being congratulated from the college, but there was no such message for me! The place was full of memories of my grandparents' successful lives about which I had learnt in many conversations and my grandmother who had died there just a couple of years before, and of occasional encounters there with famous scientists including Rutherford and the Russian, Kapitsa, who had gone back to the Soviet Union.

It is interesting to look back on what living in those inter- war years meant to the young people of the time, as well as to quite a number of old soldiers who had eyes to see and were not easily infected by nationalism and its hate propaganda. The cynicism of one of my contemporaries, a popular sports hero, at Oundle in our last year at the school surprised me: "We are just being trained up as cannon fodder for the next war", he said in 1931. I did not see things as badly as that, although I had already developed a Quaker-like antipathy to war and war- preparations. Seeing the tensions created by the blind politicians of the 1920s, Clemenceau, Poincaré and Vansittart in England, dominated by pursuit of revenge and the punishment of whole nations, including the younger generations who could not be blamed for the 1914-18 war, inclined one to despair.

Despite my father's hopeful view of the Stresemann Government in Germany in the mid nineteen-twenties, there were one or two dinner parties, evidently in the late 'twenties, when I was just old enough to be invited to sit at the table with my father's guests that were reunions with people - most of them, I believe, scientists or engineers in academic life - who had been with him in Gallipoli or Portsmouth during the 1914-18 war. And there was much talk around the table about "the Hun". I was - to use later slang - turned off and horrified at the lack of any sign of progress towards peace and reconciliation.

And now, at the end of the century, once again we find that the politicians of this day have lost the vision that the best of their forerunners had, the vision in 1945 that Churchill, Monet and Adenauer conveyed of a united, pacified Europe, the nations living at ease with each other and developing the rich texture of their varied cultures.

CHAPTER 4. CAMBRIDGE UNDERGRADUATE YEARS

It seems a shame that my grandmother in Cambridge died a couple of years before I came on the scene there as an undergraduate. We had been great friends, and I had shocked the family by the hilarious fun that we had had when, on days of family visits, I wheeled her in her bathchair along the cinder path from Grange Road to Coton village after she had already had several strokes. I was around 12 to 13 years old, and we weaved in and out of the bramble bushes beside the path with much laughter. My parents were inclined to be shocked, but granny reported what fun we had had. It was some years later, when I was in Cambridge as an undergraduate, that I was invited to turn up each Sunday about 12.30 p.m. to have my Sunday dinner with grandfather Lamb and Aunt Helen and so came to know them both better. I had further opportunities to improve my contacts with my aunt in the late nineteen-twenties, when my parents started the habit of going for a week or ten days Easter holiday on the continent beside either the Swiss or Italian lakes. I was never taken with them. Instead, I came to enjoy these Easter breaks quietly with my father's sister in the Sussex Weald, at the foot of the South Downs at Steyning, and her interesting conversation touching on the history of the countryside - which was a new topic to me at that time - and such things as Stone Age archaeology.

Sadly, I have to recognize that my undergraduate career at Cambridge was the low point of my academic career. After an enviable history of successes at my senior school (Oundle), including form prizes in most years and a couple of Headmaster's prizes, at Cambridge I performed far less well. There were various reasons for this. It began with my father trying to make me concentrate on mathematics or, at the least, to take maths as one of the three sciences which I would have to choose if I read for the natural sciences tripos. But my shock, and fury, four years earlier of being forced at Oundle to abandon the subjects that were then interesting me most in favour of science was still too fresh in my experience for me to accept such plans. I was agreeable to studying chemistry and physics, but the choice of a third subject was difficult and in the end I settled on mineralogy where the lecturers were renowned specialists in crystal structure. But the subject did not interest me, and they failed to inspire me, possibly because they were of rather dismal appearance and one was known to be a strong communist.

I followed my father's request not to join the boat club. Anyway, I had not the physique for an oarsman. I did, however, one July in one of the short Long Vacation terms, cox a Trinity boat to victory in a race on the river at Bedford. But the river had its lure for me, and I spent a lot of time -

Cambridge Undergraduate Years

doubtless too much - in canoes and occasional punting parties on it, tempted and persuaded by the good luck that my undergraduate years coincided with three very fine, sunny summers in 1933, 1934 and 1935.

My father also forbade me to join the Cambridge University Union, the great debating society. To compensate myself, being strongly interested in politics, I sampled once or twice the meetings of both the Conservative and Socialist societies at the University. And so it happened that I witnessed Burgess and Maclean performing at the Socialist Society some twenty years before they both defected to Russia. I was not attracted to them nor to either of these clubs of those committed to the political Left or Right. My interests were rather in politics with a social democratic and Liberal flavour and in religion understood through Scottish Presbyterian and Quaker ideas and practices. Two people in particular influenced my thinking. One was Dr. Alec Wood, my physics lecturer, a very well-known character at that time, whose lectures were enlivened with a great fund of humorous illustrations, stories always told with a twinkle in his eye. He was an elder of St. Columba's Presbyterian Church in Cambridge so I met him quite often also at the Sunday afternoon tea parties regularly held in the manse. He was also an active Labour party member well known as a pacifist and a city councillor. The other person who influenced me notably at that time was the minister of St. Columba's, the Revd. George Barclay. Although neither of these men made the slightest effort to proselytize me - and there was no need in any case - their influence was strongly resented by my father. I fear they may have had to suffer from false accusations and verbal attack, if only through the mail.

Insofar as the history of the Reformation in Scotland, and the accompanying establishment of democratic government in the Church of Scotland, set up the most effective safeguards against the dictatorial claims of an authoritarian priesthood, I became fascinated with the Scottish development and wanted for a time to be trained for the ministry of the Presbyterian Church. My information came very largely from the writings of Sir Walter Scott and Robert Burns. Ultimately, it was two good friends of our whole family, neighbours in the Hampstead Garden Suburb, Jock Elliott, the family's lawyer and his wife, Meg, who persuaded me to drop the idea of going for the Presbyterian ministry because my father could not stand it. This was strange, because he had occasionally expressed approval of - not to say admiration for - radio sermons and talks by two Scottish church ministers, the Rev. George Macleod, founder of the Iona community, and a Rev. Oswald Milligan, minister of Corstorphine parish church, Edinburgh. But I supposed the intensity of my father's conviction of his rights and duty to guide me prevented him seeing his inconsistency over this.

Cambridge Undergraduate Years

There were other curious sequels. It was through the Elliotts that I first came to know James Paton, whose wife Ginge was Meg Elliott's sister. Jimmy Paton was a meteorologist and aurora specialist in the Department of Natural Philosophy in Edinburgh University. He and his family also became firm friends of mine. And, by a strange chance, years later, I met and married the Rev. Oswald Milligan's only daughter, Moira.

Other people suggested by my father as models to inspire me were some of those great religious leaders who had shown, or were showing, a strong humanitarian concern: the Anglican Archbishop Wilberforce who led the so largely successful campaign for the abolition of slavery in the early nineteenth century, and Albert Schweizer, the German Lutheran missionary doctoring in the first half of the twentieth century in West Africa, who raised money for his own mission, hospital building etc., by his organ recitals in Europe. I thoroughly agreed in honouring these choices, but as time went on I was naturally more inclined to make my own choices. Characters who particularly impressed me in the nineteen- thirties were Mahatma Gandhi with his non-violent crusades and Pastor Martin Niemöller, the former U-boat commander from the First World War, who became a staunch leader of the German (anti-Nazi) Confessional Church. He came and preached in the United Free Church near our house in the Hampstead Garden Suburb one Sunday. And there were two famous Norwegian writers in the time of my youth who particularly interested and delighted me. One was Knut Hamsun, who-in his "*Markens Grøde*" (the growth of the field) portrayed the characters whose hard lives and struggles opened up remote land in central and northern Norway for tillage, winning it from the forest. The other one was the Catholic authoress, Sigrid Undset, whose great historical novel, "*Kristin Lavransdatter*", (in English as well as in Norwegian) illustrated life in medieval Norway, including business with Iceland, through the private and public affairs of a landed family and typical cloisters in south-eastern and central Norway, as far north as the Trondheim district from the prosperous time in the late 1200s to the arrival and early spread of the great plague, the Black Death, in 1349-50.

In my Cambridge undergraduate years I had no serious girl-friend, but I did have quite a number of very pleasant female friends. It turned out that most of them were either auburn haired or blond or the type that is known in Scandinavia as mørk-blond (dark blond), and it seemed that the attraction in these cases was usually mutual. There was, however, an amusing episode connected with the chemistry lectures which I was attending. One darker haired girl in the class was undoubtedly a striking beauty, and many of the men students wanted to "strike up a friendship" with her, though none seemed to be able to. One worked out a plan to fall off

Cambridge Undergraduate Years

his bicycle close in front of her, so that she would fall over him in the chaotic rush when the lecture ended. But when he put the plan into operation, falling very close in front, the lady steered round the obstacle faultlessly and rode off with her head in the air.

My most notable relaxation in my undergraduate years was visiting my school friend from Oundle, Mike Scratton, who went up to Trinity Hall in the same year as I went to Trinity, and his widowed mother and sister Deborah. She was still at school and younger than my other friends but always friendly. Mrs. Scratton was always urging me to go in for a political life. Her husband had been an army officer. We last met when I was on home leave late in the war, and she told me: "You know, after the war, all the votes will be Labour". And that was in those days the general assumption of all young people of gentle, liberal outlook. She shared it too. But another important aspect of her household's influence on me was the continuous succession of young Scandinavian - mostly Swedish and Finnish - girls that Mrs. Scratton had staying, usually for about a year, with her, to perfect their English. They were, I think, all nicely brought up and very attractive girls. Mike Scratton himself married one of them, though she sadly died after a year or two, having given birth to a sweet little daughter. It was a new experience to me at that time to listen to the 3 and 4-language conversations between these girls and their friends. Each would speak in their own language (or in Swedish in the case of a Finn) and each would reply in her own language and be understood.

At the beginning of one autumn term, in October, I went one Sunday afternoon to call at the house about 2.15 p.m. and found Mrs. Scratton and Deborah so far alone. Mrs. Scratton asked me at once "Have you had tea? Won't you come in and have a cup?". I pointed out that I had not long finished my Sunday dinner - that was before the days when the English started calling every meal eaten in the middle of the day lunch, however substantial it might be. The kind Mrs. Scratton gasped and said: "Oh we thought it was a quarter past four. We must have changed our clock the wrong way". It was the day when the clocks should revert to Greenwich time! Although the day was half over, she and Deborah alone in the house had not discovered that they were wrong by two hours!

I had other friends too, of course, in those Cambridge days. Peter Leggett (actually he was D.M.A. Leggett), who had the room above mine in Whewell's Court, Trinity, was a particularly close friend. He was reading Maths, as were two others of my best friends there, Carl Marsden and Archie Black. The former returned after his degree to his father's lawyers' firm in Blackburn, Lancashire. But Archie became a professor of engineering at Southampton. Peter Leggett, however, undertook an essential

Cambridge Undergraduate Years

service to me. He loyally woke me every morning and came down a second time some minutes later to check that I had stayed awake. Once he suspected that things were really too quiet beneath his room, even though he had had to roll me out of bed to get me to wake on his first visit. He then found me sleeping deeply on the wooden floor underneath my mattress.

Peter and I and Carl Marsden went on holiday together in the summer vacation in 1933 and again in 1934. Peter became a lecturer in mathematics first at Queen Mary College until 1939 where my father still was, then went to the Royal Aircraft Establishment until 1945, leaving to become Secretary of the Royal Aeronautical Society - a sudden break in his career which had to do with his revulsion at the obliteration bombing of the German city of Dresden. Later he was to become the first vice-chancellor of the University of Surrey in Guildford. In 1933 my summer holiday began by riding my little two-stroke motor bicycle from my home in north-west London to his home on a hillside just out of Biggar, Lanarkshire in the Scottish Borders. On the way I spent a night with my old great aunt Annie at her house in Darlington and, on arrival, was brought a big jug of boiling water and an old hip bath on the straw mat in my bedroom. It was a great thing to ease my stiffness, but once in the steep-sided bath tub my very stiff legs were stuck until the stiffness eased. After a couple of nights at Peter's mother's house, we were joined by Carl Marsden and all three of us set out again, on hired push-bicycles, for a long trek through the Scottish Highlands. On the way we night-stopped in youth hostels and broke the journey for one or two climbs. The first was the giant, very steep, grassy cone of Ben More (about 3,850 feet/1170 metres) near Crianlarich, and I learnt at once that my smooth-soled shoes were not good for that sort of work. I had a much more alarming experience from the same cause a couple of days later when we were making our way sideways along another steep grassy slope farther on towards the north-west, somewhere above Glen Shiel and Loch Duich with a cliff edge a few metres below us. Again my shoes slipped, and I was soon on my back, sliding down the slope, with the precipice only a few yards away. Without time to think I got myself rolling fast sideways along the slope and soon reached a point where there was no longer any sharp drop. I am ashamed now to think how many times I made the same dangerous mistake of having only inadequate shoes. It was probably the result of trying to travel light. In later life, to correct this, I have often taken too much to carry with any ease. At that time, we were very inexperienced travellers, and we had other experiences, some of them pleasant surprises. Despite the generally very fine summer, we did have one long cycle ride in very heavy and continuous rain. We were in the Western Highlands after all at that time. All the way up Glen Garry we got wetter and wetter, and when we reached the lonely hotel at Tomdoun we must all three have looked like drowned rats. It was late for tea and some of the tables showed the signs of having been used by a large party. We wandered in and asked the youngish manageress if we could possibly get tea. She looked sympathetically at us and our sodden clothes and answered our inquiry by waving us to the nearest table, which she then replenished with good Highland scones, oat cakes and bannocks as well as pancakes, butter and jam. We thanked her and asked the price and were amazed when she said it would be one shilling and sixpence each. We then got back to our saddles, and plugged on over the hill, till freewheeling down through Glen Shiel to the sea loch, Loch Duich, we got a splendid view of Skye with the weather beginning to clear, and it took our attention so much that two of the bikes nearly collided, at speed!

We went on to climb the Cuillin Hills in Skye, where the hardness of the old volcanic rock tore our shoes to pieces. And then we rode on to the northern tip of the 50-miles long island, beyond Uig to Duntulm. There, there was a boat - quite a heavy boat unfortunately - that had been pulled up from the beach on to the heather. With the beautiful cloudless, calm weather it was a tempting sight. So we walked to the only house near and asked if we could possibly hire the boat for an hour or so. The young man we asked replied that we could certainly *borrow* his father's boat. Of hiring it there was no question. So we did, and rowed out to the island, and I had a swim from the stern of the boat .... but what a cold swim! The very clear water looked wonderful, but it was not as warm and inviting as it looked. Another snag was the tremendous effort involved in dragging the heavy boat up the beach and over the heather. At the end of it we were, all three, exhausted. Then Peter discovered that his pocket watch was missing. And it was a treasured memento of his father, who had died the year before. So we set about searching the heather, particularly where we had had the hardest effort pulling the boat and where there was a dry ditch which I remembered we had jumped across on our way down to the boat. After more than half an hour we had to give up the search, and poor Peter was understandably sad. I had been standing waist-deep in the ditch and had to put down my hands deep into the heather beside it to haul myself up. As I did so, I felt the palm of my hand close around the watch (which at first I still did not see)!

The next year the same trio of us went to Switzerland in the summer vacation. I had always found the German language easier than French, especially the spoken language being usually more slowly and clearly spoken, so we went to German-speaking Switzerland, arriving by train one really balmy summer evening - a real continental warm, still summer evening, with a fragrance in the air from flowers and trees that was

Cambridge Undergraduate Years

a new experience for me. We walked up through the city to the youth hostel. Later we went on by train to Andermatt, up in the Alps towards the Gotthard pass, and then walked the odd mile on to Hospenthal where the road divided and there was a youth hostel.

First, we had stopped off for a few days in the Black Forest, to sample a little of Germany itself. We climbed to its highest point, Feldberg, after a night stop in a youth hostel somewhere on the high ground and were subjected to German discipline - everybody up at 6 a.m. and a cold wash under a tap outside. It was scarcely a month after the massacre of some supposedly leading figures in the Nazi party in the "Night of the Long Knives", but we were of course not introduced to any of the bad or suspect sides of Nazi-ism. The noticeable features were the flags everywhere and the anxiety of any Germans one spoke with to explain and justify themselves - certainly evidence of a feeling of isolation from the European community and that Germany had been badly treated. I remember a graceful 1914-18 war memorial in the form of a stag outside a big monastery church, inscribed in German "Bold as a stag in the forest, Germany stands against the world", expressing this isolation.

Our progress in the Alps in Switzerland began with a long, long days walk in the heat, first over the Furka pass, down the other side with views of the Rhone glacier, far retreated from its nineteenth century positions which reached the valley bottom. The descent wore out the seat of my scouting shorts. Down in the bottom we were able to take a short ride on the Zermatt train, just five to seven kilometres, to lead us to the bottom of the Saas valley, up which we headed on foot along the narrow, dusty mule track that was then the only highway in the dale. At a farm at Saas-Balen we decided we could go no farther. Our feet had taken us about 26 miles, the first, but also the longest day's trek of the whole holiday. (Another mistake which I have perpetrated on too many occasions. Exhaustion comes sooner when one is out of practice.) But we got permission to sleep in a barn beside the track, to which our attention was drawn by an electric light on the outside. Part of it was a hay barn, but we made our meal in a safe place and had a bottle of red wine that we had bought along the way with it. That was all too much for me and I spent a miserable night of sickness: my first realization that, like many people, I have to be especially cautious with red wine. However, we pushed on the next day. The track became steeper and went up past a succession of little shrines with religious figures painted in strong colours. There were pilgrims gathering at some of them, because it was Maria-Himmelfahrt (Mary goes to heaven) day. Before that day was out we reached Saas-Fee, which stands quite high on the side of the valley. It is now reached by a broad, well-made road, and we booked ourselves in to the

Cambridge Undergraduate Years

youth hostel where we had to sleep in an old log-cabin type hut directly above two or three cows with the bells still attached to their necks. It was not a quiet night and a bit smelly, but for most of the night even I slept the sleep of the exhausted.

During one of the days of our stay in Saas, we managed to climb to the top of the Mittaghorn (3,148 m. or about 10,330 feet), which stands out toward the middle of the valley, about 3 km south of Saas-Fee, and rather nearer Saas Almagel. South of it, two of the glaciers sometimes reach the valley bottom, blocking the stream and so forming a temporary lake, the Mattmarksee, until the ice gives way sometimes producing a flood which rushes on down the valley causing considerable damage to buildings and whole villages. Between 1589 and 1772 such disasters occurred there at least a dozen times, and then the lake became a normal feature until the late nineteenth century. Other Alpine valleys experienced parallel cases. And there were two or three minor recurrences of the Mattmarksee and the collapsing ice dam in Saas between 1965 and 1970.

For our climb, however, in August 1934 we had nothing but cloudless weather and great heat. By the time we came down Peter and I seemed to be suffering from a touch of sunstroke, and on getting back to the youth hostel we had a craving for tomatoes with salt and a drink of water, after which we gradually felt better. While there in Saas Fee a few days later we had another experience, when I had decided to make some scones in the hostel. When we had roared up the wood fire, and I had made and cut the dough ready to put into the oven in the cooking stove, I stooped down to put the material in. But the moment I opened the oven door I was overcome with the pungent vapour belching out from the pine sticks that someone unknown had filled the oven with, presumably to dry. And that was the end of our baking project for that day. Indeed, it was nearly a year before I regained my love of the smell of pines and pine smoke.

The next year, as things turned out, was to be the final year of my undergraduate career in Cambridge. Chemistry and mechanics were the subjects which I was having most success with in my studies. I did not want to become an industrial chemist, however interesting the processes involved in making new and valuable materials might be. And I certainly did not have the skills needed to become any sort of engineer. So, clearly, some redirection of my training was called for. At this point I was relieved to find that my father's older sister, Aunt Helen, was willing to advise me, though my father did not like her advice. Moreover, she was on the spot in Cambridge, easy for me to see, and she knew the university, its people and facilities. Her opinion was that I was the sort of person who was well suited to the geography department. Though the subject was not universally well

Cambridge Undergraduate Years

thought of, the staff was very good and there was a wide variety of interest in the courses. Unfortunately, the normal requirement for the first part of the tripos examination was a two year course, and my father would not hear of me staying on for a fourth year as that would imply. Far less was there any chance of a fifth year in which to complete the second part of the geography degree. The upshot was that I did change over to the Geography School, under Professor Debenham, who had been south with Scott. But, although I worked far harder than I had ever done before, read very widely for the course, and became very interested in the physical as well as the human geography and historical geography studies, I only got a Second Class Honours degree in Geography Part I. And my father told me that I had "only obtained a mediocre, mixed degree that I would regret all my life".

CHAPTER 5. AFTER THE DEGREE EXAMINATIONS

Before coming face to face with the alarming prospects ahead of me, however, I had a very nice invitation which also influenced my future. At the end of the 1934 holiday experience of central Europe, I was clear that I wanted next to see Norway, where the sea as well as the mountains are an integral part of the picture. Also I felt the need to fill out my acquaintance that was already begun with the treasured book of old folk tales and Sigrid Undset's very realistic portrayal of medieval Norway in *Kristin Lavransdatter* with a look at the country in our own times. I therefore announced to my friends that that was where I would go next, at the end of the summer term in which I took the degree examinations in 1935. Another of my Trinity College friends, a Quaker, Dennis Conolly, decided to come with me, and coupled it with an invitation from his parents to go on to join them and Dennis's sister Ann in Sweden afterwards. Dennis was visiting friends in Germany that July, and so we arranged to meet on the quayside in Kristiansand, near the southern tip of Norway. I began to learn the Norwegian language from the little books of Hugo's "*Norwegian in three months*" and soon was making encouraging progress with it. Moreover, after trying many people's helpful suggestions and finding all the regular passenger lines expensive and already booked up, I managed to get a berth for about £12 on an Ellerman Wilson Line cargo ship, which was loading sugar at the Tower Wharf dockside in the Pool of London for shipment to Kristiansand. Arranging this was a difficult operation, which also entailed several telephone calls to Dennis in Germany keeping him informed about the delays to my departure from England, finally amounting to six days till I went aboard (seven days till we sailed on July 22nd) and gratitude to our neighbours, the Elliotts, in Hampstead Garden Suburb who produced extra meals to help day to day to keep me fed until I really got away. The ship rolled quite a bit in a north-westerly swell in the latter half of the crossing, although the waves were only 3 to 4 metres high and the weather had been mainly fine and warm. No-one was sea-sick but on going ashore next morning, I fell in the middle of the stone quay which seemed to be heaving up under me as the sea had done. Then I met Dennis on 25 July 1935, without a hitch, although arranging the meeting had seemed difficult, since neither of us had ever been in the country before nor knew any landmarks in the town.

Dennis and I spent two weeks wandering through south Norway, starting by local train and then bus up the great valley Setesdal that led us up due north into the heart of the mountains. We spent the nights in youth hostels, where we were sometimes served at meals by maids in uniform and white lacy caps and enjoyed their fish, followed by strawberries and cream. At some of these hostels the coffee (excellent quality) was accompanied by cream and sugar served on a little tray with dainty traditional silver and glass vessels - an old-fashioned elegance that has since sadly disappeared. At one hostel, which was a quite small farm cottage, there was an unforgettable loo in the form of a bridge over the rushing stream at the bottom of the little garden. One climbed the fairly steep rustic steps to enter the loo which was a sort of log hut-bridge across the water about 2.5 or 3 metres (8 to 10 feet) below. And there was the row of four round holes, side by side, with the old agricultural catalogue at one end to provide the paper, a style possibly owing something to the American Middle West.

At that time, there was no road through to the north from the top of Setesdal at Hovden. So we had to cross the mountains. We had a map and took the path for Haukelisaeter, 3 Norwegian miles (i.e. 30 kilometres or about 20 English miles) on. The weather looked sickly and soon began to rain. The people at a hut encouraged us by telling us that the path ahead was well marked by cairns. We needed the encouragement, because our map told us that the surveying had been done in 1853 and the detail could not be guaranteed. We soon discovered what that meant. Many of the mountains were not marked at all, and the path led us on the wrong side of several lakes. Around midday we came across a big herd of maybe 200 reindeer. After a lunch picnic we got very wet wading a stream and came upon an unmarked hut in which there were scribbled notes of gratitude, written by parties from three different places over many years who had been benighted there. There was, however, no mention of Haukelisaeter, which we were making for. In the afternoon we came across the first patch of snow and soon had to cross a good deal more of it. Then we were in the mist, using our compass and concentrating our whole being on spotting the next red-painted cairn ahead. The cairns were good and frequent, seldom more than fifty yards apart. By 5 p.m. we were feeling rather miserable and wet, but at 5.30 p.m. we came to a signpost, where the path branched. The path was still wrongly shown on the map. But at 7.15 p.m., wet to the skin, we arrived at Haukelisaeter on the road over the great hills from Telemark to Hardanger and the fjords in the west. We found this large wooden hostelry some 1,000 metres above sea level and, although it was rather expensive, booked ourselves in and entered the lofty dining room, itself a great chalet/hut in the Telemark style, and sat down at one of the long tables for a very welcome meal. I found myself able to ask in Norwegian for the various dishes to be passed along the table. But we also got talking to an English nobleman, in very Highland style fishing rig, who habitually came for two months in the summer and made mine host uncomfortable with his patronizing air.

The next day was a Sunday but we were up in time to climb into the "*melkebil*" - a typical American farm truck with an extra seat behind the driver operated as a bus for a few passengers and (mainly) to collect the milk for the cheesery. It took us over a high, cold pass between the mountains and then down steeply to near sea level, passing two thundering waterfalls and great rock-slides and scree on the way, to a little place in the bottom called Røldal. There we recognized (too late!) the cheesery by the powerful smell of goats' milk and cheeses that was filling the valley. Our vehicle had been backed right into the building that was the source of the smell before we recognized what we were in for. At the Røldal Hotel nearby we had a good, cheap breakfast and got into dry clothes. But the call to the bus came before we had really finished. We reached Hardanger fjord at Odda at 12.15 p.m. and sailed all down the fjord from one little port to the next one, to reach Bergen at 6.30 a.m. next day. Ripe cherries were brought aboard by the passengers who came on at each place. I succeeded in having a lot of fun with some children for two hours on the deck, telling them successfully enough in Norwegian that if they ate too many cherries I would throw them into the waves, and that at 9 p.m. each night I become a troll and will have three children to my supper. This game went on intermittently, keeping all the players amused, for about two hours. It was an older girl (and very pretty) who got on board that evening who told me of Knut Hamsun and encouraged me to read some of his work.

Making our way further up the coast and round into the fjords and valleys, we came to Voss, a popular beauty spot. We looked around for the youth hostel and found what we thought was one with breakfast being served in the "people's high school" (Folkehøgskule is the country spelling of the language, Landsmål, much in evidence there in the west). People were queuing to go in to the breakfast, so we joined the queue too and paid our 75 øre for it. But nobody checked our youth hostels card and we concluded that we had unintentionally "muscled in" on a conference organization, since it was evidently a gathering of young people (up to their thirties) from all over Norway. Anyhow, we paid for our food, after sleeping on the floor, and no-one was any the worse for our intrusion!

We hiked part of the way up the beautiful valley to the summit at Stalheim, which had been a favourite holiday spot for the Kaiser, whose visits are commemorated in the name given to the viewpoint above the hotel: "Kaiser Wilhelms Høi" (= height). And from there we went on down the long, steep-sided narrow valley, Naerødalen, with very high waterfalls spilling over the cliffs from the top edge on either side. And we were

After the Degree Examinations

puzzled and amazed by the ancient (*stolkjaere*) horse carriages (without their horses) hidden behind the rocks at every turn of the road. The explanation became obvious next morning after we had spent the night in one of the little white wooden cottages that was the youth hostel, near the quay in Gudvangen, an amazingly scenic place, still hidden away at the foot of the great cliffs on either side of that valley and at the head end of the little fjord that fills the lower end of the dale. A Belgian cruise liner had come in overnight, and the ladies and gentlemen travellers, dressed as if for the Champs Elysées in Paris, were coming ashore in the bright morning sun and were all being helped up into the little horse carriages (with their horses present and harnessed now).

Our fjord boat worked its way down the narrow, winding Naerøfjord out into the wide Sognefjord, the biggest of all the fjords, at Balestrand with its posh hotel, where Dennis and I got a row from the ladies sitting in the blazing sun behind glass doors, because we opened the doors to go in. We came back along the Sognefjord to go into another branch fjord, Aurlandsfjord. We were lucky to be gliding our way on the water through such wonderful scenery in cloudless sunshine. But that evidently did not affect a party of French ladies who stayed below decks all the time, playing bridge! At Flåm we got off the boat and spent a warm afternoon slogging our way up the steep zigzagging road (twenty-one zig-zags!) to reach the mainline Bergen to Oslo railway at Myrdal station, at the top. The (Vatnehalsen) hotel did not open until after our arrival at 10.20 p.m.. But the pretty daughter of the proprietress was amused by our efforts to talk Norwegian and gave us a tremendous spread of an evening meal for just 2 Kroner each.

At half past eleven we retraced our steps to the station and bought our tickets for the midnight train to Oslo. We got the last two seats in a compartment, but it was not a restful night with the other passengers keeping the windows shut, despite the warmth of the weather. The man sitting next to me patted me on the knee often enough to keep me awake while he prattled on, speaking fast, in toothless Norwegian and with much laughter.

We found the youth hostel on the edge of Oslo, in a grand modern building Den Lutherske Hotell, had a good lunch of mackerel hash followed by blueberry porridge and a good long sleep. The Norwegians make a fruit porridge (fruktgrot) with a little potato flour to thicken the juice of any fruit. To me it makes a tasty, refreshing sweet whatever the fruit (or rhubarb) that is used.

While in Oslo we behaved like normal tourists, seeing the sights and taking the "Underground" that has an extension to the top of a

After the Degree Examinations

mountain (1,700 feet above sea level or over 500 metres up) in the nearby terrain of Nordmarka, which is one of Oslo's glories for its own inhabitants. We then moved on with the mainline train to Stockholm and met Dennis's parents and his sister Ann, who was still sometimes wearing her calliper to strengthen her leg that had been affected by polio. More sightseeing, including the fine modern city hall with its tower at the water's edge, and a visit to a big international scout jamboree, before we went south by train to Gränna, the pleasant country town beside Sweden's second biggest lake. On the way, we made a memorable excursion in order to sample the famous Göta canal and the little steamship plying along it. This involved leaving the mainline Stockholm to Malmö railway at Linköping and taking to a local line for 10 km. and finally a car to reach a village called Berg near the canal. All this in a charmingly rural countryside and along leafy lanes. But the changes were rendered incredibly complicated by having to transfer our party's total of twenty-one cases and parcels, including a rolled up home-weave stair (or passage) carpet, passing each item from hand to hand and seeing it properly installed in the next vehicle. I evidently fully shared the anxiety of seeing the transfer completed without loss, because it came back to me in repeated nightmares over the next five to ten years. Our party slept the night in a little country hotel (which we just filled) in the village: the three of us males had beds in one big room, the ladies had beds in another. The furnishings were old fashioned, with several little tables each covered with a loose lace cloth. I probably put the contents of my pockets out on the cloth on the table next to my bed, and when I turned over in the night everything fell. The lady of the hotel woke us at 6.45 a.m., but shook her head and said it was too early for breakfast. So we had to have it on the ship, which we boarded at 7.30 a.m. at a point where the canal intersected a country road at an iron bridge that was rolled back by hand. At the breakfast table, at 7.45 a.m., I discovered that my passport was missing. As we were going on to leave Sweden for Denmark later that day without any coming back this way, the passport had to be found and fetched at once, while the rest of our party got on with their breakfast. There was no difficulty about getting off the slow-moving boat at another rolling bridge where the canal was always very narrow. Then I ran back to the village along the towpath and found out that it was becoming a very warm day. Breathlessly I explained myself at the hotel, and my passport and some other papers were quickly found on the floor at the back under my bed, where they had slid to, after falling in the night. Very quickly I was able to commandeer a taxi, and the taxi-man obviously enjoyed plotting a course to a good point on another road that intersected the canal. So we drove a mile or two along some more pretty lanes in the sunshine and arrived at a bridge just as the canal boat was approaching. I was quite dismayed when it turned out that the boat was not going to stop, but I safely grasped the thick rope that was put over the side for me and clambered up on to the deck a few metres above the road. And so to resume my breakfast amid general amusement!

I see from my diary how I was dizzied in every town by all the pretty blond girls along our way on this holiday and amazed at how approachable many of them were, often starting conversations themselves. There was an innocence that merged into a happy carelessness about the sexes in both Norway and Sweden that was in those days quite unlike Britain. And bathing from the sandy shores in the great lake Vättern[4] in Sweden, many of the younger teenage girls were very inadequately covered. The children however, were really beautifully turned out, particularly on Sundays, the girls in light coloured dresses and always with a great big bow on the top of their heads.

We stayed a few days in Gränna in a hotel where we ate our meals in the sunshine in a lovely rose garden with a white picket fence. We enjoyed bathing in the lake to which I returned many years afterwards, camping with my family, and the lovely sunset scenes under the trees and across the water. We also re-visited the cake shop on the corner, where the "Napoleon cakes" - cream slices with one layer of vanilla custard and one of cream between flaky pastry - are an experience not to be missed. And then I accompanied the Conollys further for a few days up in the rolling hills a little way to the east of Gränna at a hotel that specializes in convalescents, at Örserum beside a little lake, a quiet spot where we found places for sketching the peaceful landscape of birches, rocks, lakes and heather.

Even more important not to miss is a trip to the big island, Visingsö where a horse-drawn charabanc took us through the pines and up the hill to a little medieval church with the interior walls covered with figures lightly painted in red and then down to a much bigger, wooden seventeenth century church with a steeple tower and a predominantly blue colouring that went well with its setting in a cleared space among the pines.

Then we went on by rail and ferry to Malmö and Copenhagen, where I saw the Conollys off on their way home. That afternoon I called on those old friends of the Lamb family, the Fluxes, who greeted me affectionately and took me by tram to dine at the Langelinie pavilion and walk through the long garden beside the harbour. First, we passed the powerful allegorical sculpture of the legendary team of bulls ploughing Denmark out of southern Sweden with water springing widely and steam belching from their nostrils. The path leads on to the justifiably famous

---

[4] Vättern is about 70 English miles long by 9 to 10 miles wide!

Little Mermaid (*Den lille Havfrue*) statue, showing the beautiful girl - modelled on a much loved nineteenth century Copenhagen actress - with her fish tail, sitting up on her smooth rock at the water's edge. And then, behind it all, we come to the busy harbour shipping and the lights twinkling in the peaceful dusk above the water. The sky was clear, but the light nights (when the twilight is continuous) had ended on August 8th. Back at the Fluxes' flat we drank toasts to my family, my holiday and my coming attempt on the Consular Service entry examination - which is now the plan ahead.

I had already bought the tickets in England for my journey home on the Danish railways and the North Sea ferry from Esbjerg to Harwich. But they were dated ten days on. So my immediate problem was how to live those ten days on the roughly thirty shillings (nominally £1.50 in modern coinage) that were all the money I had left. (To be precise, I had 33 Kroner in the Danish money of that time.) Solving that problem was an instructive and very well worthwhile exercise. But I was determined to see as much of Denmark and Danish life as I could on this, my first visit. Staying in the youth hostels, first in Copenhagen, and later in country towns in Jutland, my night stops cost me only one krone, and in order to spin out the money I breakfasted as late as 11.30 a.m. At the "au Tomaten" coin-in-the-slot place in Copenhagen the "*morgen komplet*" (breakfast tray) cost me less than 1 krone, including a cup of good strong coffee. In fact, the outlay was even less, because I already had the *knekkebrød* (Ryvita equivalent) in my rucksack. By evening, after walking about the city or the country town, I was very hungry and found it a strain to have quite poor Danish people sitting beside me smoking cigars while I could not even afford a hot sausage! In Vejle, in eastern Jutland, I bought bread, butter and honey, all nourishing and cheap. I walked a mile or more out of town along the edge of the peaceful lowland fjord, lay in the sun and began to study an economics textbook. That evening, with half of the ten days behind me, I worked out that my money was lasting out and that I could afford to make myself a more substantial meal. So I bought a fair pile of cabbage and beans and cooked it for my supper. It was not the wisest choice, and I regretted it during a night of diarrhoea! However, later on the next day, I was buying eggs and cheese and good, plain cakes, and was quickly recovering. And in the days after that I improved further with half a melon, which cost me the equivalent of sixpence, and mild Dutch Gouda cheese. On the last day in Denmark I even had eggs and bacon with my coffee.

In another country town, Kolding, on the Sunday afternoon I met a couple of youngish middle-aged local people. Herr and Fru Vagn Jensen, who were walking their bicycles up the long hill into the town after a ride in the afternoon sunshine. They invited me to supper with them in exchange for an English lesson. (That was difficult because I could not understand his Danish, with its "flat" pronunciation and swallowed syllables, however slowly he spoke.) But we became friends. Years later, after the war, they visited us in our first married home in England, and we paid a return visit to them with our first child. Unhappily, that was not a success: the Jensens, being themselves a childless couple, had hired a young girl as baby-minder to take charge while we went out in our car. This plan was vigorously opposed by our one-and-a-half-year-old, who saw in it a plot to get rid of her!

CHAPTER 6. TOWARDS A DEAD END

It was now decided that I should enter for the Consular Service examination in 1936. To this end, my father used his knowledge of the University of London to arrange for me to attend lecture courses at University College in economics and enough modern languages of which I already had some knowledge, namely French, German and Norwegian. I spent Christmas 1935 staying for three weeks in Paris with a French Protestant family from the south. There were other students as guests with the same family, three young Americans, one a man and two girls, already speaking good French as well as one untypical Swedish girl. But it was an ominously ill-fated venture. Madame Sarrut and her son were very nice, and made us all feel at home with plenty of Christmas festivity, including a pudding over which so much lighted brandy was poured that the jug caught fire. What spoilt the hoped-for success of the visit had most to do with the awful weather, continuously dark and grey and raining, with the river Seine flooding, and my too limited ability with the French language. The minister at the Sarrut's church, where I attended the main Christmas service, had spent much time in London. The church building had been Catholic until the French revolution. I also attended a picturesque service for Swedish visitors at an Alsatian Lutheran church, which was packed for the occasion, with a great Christmas tree, topped by three stars, and surrounded by white-clad peasant girls in the softly lit apse. But it came out afterwards that Mme Sarrut suspected that the Alsatian community in Paris was deeply infected with Hitlerite influence and activities, and she may have been right as the French state had been using its opportunity since the victorious end to the 1914-18 war to reduce or even eradicate German sentiment and traditions in what had historically been a state with German language and traditions. The Christmas Day service at the (Protestant) Église de l'Oratoire was attended by a massive congregation of two thousand, filling the church, which included some of the obviously richest as well as some of the plainly poorest people I had seen in Paris.

At the weekend after Christmas I went up to Montmartre sight-seeing, but the huge white domed church struck me as more like a Mediterranean mosque than a church; and, inside, the crowded congregation, periodically mouthing "Jésu", produced an almost rhythmic sound like regular little gusts of wind. The atmosphere in the great church with many people arriving and going all the time, was to me suggestive of a railway station, and did not appeal to me.

There seemed to be an undeniable gulf between the outlook of all of us guests in the Sarrut household, the three Americans, the Swedish girl and myself, whom Mme. Sarrut saw as representing the suspect "monde Anglo-Saxon" and the French community around us. This was sad because the Sarruts were such kind and generous hosts, but it was nonetheless a real difficulty. One place where we all used to congregate, when we went out together exploring the city, was a little Danish café in an upstairs room on the Place de l'Opéra. It was elegant, and the coffee and cakes were very good. The only item of possibly questionable taste was an elderly Danish lady who seemed to be an invariable occupant of a table on one side, richly dressed, usually in white but with various gilded features. The place made a quiet refuge in the heart of the busy city. But it was unfortunate that our meetings led to repeated expressions of exasperation about French attitudes, particularly the French political frustration in the 'twenties and 'thirties at British and American-suggested efforts towards pacification through scaling down the reparations demanded of the Germans (when they were overwhelmed by the inflation disaster and its consequences). There was a universal and all-pervading attitude of suspicion and cynical disillusion in the conversation of all our French contacts and, most of all, no understanding at all of pacifism.

I tried to advance my French by reading Léon Blum's book *"Souvenirs sur l'Affaire"* about the scandal of Captain Dreyfus that had rocked France in the eighteen-nineties and the writer, Émile Zola's account of it and the part Zola's research and revelations played in it. Dreyfus was a 30-year old officer in the French army in 1889, son of a rich Jewish clothing manufacturer, posted in 1893 to the war ministry. Born in Alsace at Mulhouse (formerly Mülhausen), he fell under suspicion of passing military secrets to Germany and was arrested late in 1894. The courtmartial quickly found him guilty on flimsy evidence and in secret and he was sent to Devil's Island, Guiana, for life despite his plea of innocence which was consistently upheld by his family. But a section of the French press, the army, and public opinion at the time, infected by anti-Semitism and confused by evidence that a Major Esterhazy of the same army section was spying for Germany (though he was acquitted later) and other implications of corruption in high places produced a highly charged atmosphere. It took some years of agitation, in which at least one respected elderly politician and Georges Clemenceau as well as many writers from Anatole France and Marcel Proust to Émile Zola took part, to bring attention to the doubts and irregularities in the case, before Dreyfus was brought back to France for retrial and it was 1906 before he was fully re-instated. The nation had been divided as between defence of the civilian order of the Third Republic against the forces of the political right, including the military, which believed that the national security was threatened by Jewish intrigue and international socialism. Émile Zola himself, who had "blown the whistle" back in 1898 about the dubious nature of the first trial in an open letter to the president of the republic, had had himself to stand trial, was found guilty and imprisoned for a year when agitating for the revision, and there had been anti-Jewish riots in some provinces. Amid the passions aroused, church and state were separated in 1905.

The book was a great introduction to French political life and French institutions, but it did not really solve my problem with the French language. This had most, I believe, to do with the speed at which French people generally speak and which has meant that I have never learnt the ability to distinguish clearly or quickly enough the words in the spoken language. This was not altogether a blockage on the language, but the implications for my intended course of studies were serious.

I continued my language courses all through that winter. I was interested in a French professor's admiration of the flexibility of the English language, with its freedom to use almost any noun as a verb when it is useful to do so and will be readily understood. Many of these usages are centuries old - e.g. to condition, to engineer, to house, to lord it, to puzzle, to war against. This may indeed be a practice developed in earlier times that is becoming forgotten nowadays, when Latin and Greek endings are more widely familiar than they used to be, so that forms like "proselytize" and "revolutionize" are readily invented and need no "popularizing". The German classes caused me no great problem but I did not go far with it and never really reliably mastered the grammar. Soon it was clear that Norwegian was to be my best foreign language. I found its simple grammar encouraging, and many of the words are close to English (and Scottish dialect) forms, to which they are historically and understandably related. As with my earlier study at school of "science German", our charming elderly Norwegian professor (Grøndahl) made us learn some folk tales, including one or two that I had already encountered in English in my father's old book of Åsbjørnsen's and Moe's collection and some poems. Their pawky humour fascinated me as did the inconsequential endings to many a dramatic tale, nicely exemplified by the very short story of the "Lad who met the Devil": the boy had been going along a road eating nuts, when he came to a worm-eaten one, and just then he met the devil. He persuaded the fiend to show him how he could make himself so small that he could worm himself into anything. This got him to show off how he could even get right into a worm-hole. But once the devil was well inside, the boy blocked the hole and took the nut to a blacksmith to crack open for him. The job was surprisingly troublesome, and the smith had to take the biggest sledge-hammer to the nut. But then it went to smithereens, and the explosion destroyed the smithy itself. "I think the devil was in that nut", said the smith. "So he was", said the lad'.

But it had become obvious that my French would never let me pass for the Consular Service. So I had to give up the plan. And my affairs seemed finally to have reached the dead end.

The stresses and anxieties of that year after taking my degree examination, until I withdrew from my courses at University College in the spring of 1935 and became unemployed, were deeply depressing. My situation was alarming because of the mass unemployment at that time. My father would not hear of me applying for any of the driving jobs that I thought would be better than idleness. So I wasted a lot of time over two or three months, and the money spent on postage stamps, applying for the various jobs I found advertized in my father's *Times* newspaper. Ever since coming home from the Cambridge examinations I tried *The Times* crossword puzzle each day for just a year, when for the first time I succeeded in completing the solution. For the most part, the clues were too abstruse or too classical for me. I thereupon decided that I had wasted more than enough time on that and have never attempted a *Times* crossword puzzle since (and very, very few other puzzles either). Then, some time in the summer I spotted an advertisement in the paper for Technical Officers to be trained in a six-month training course for weather forecasting posts in the Meteorological Office. I went for the interview, and the chairman of the board before me was introduced as Sir George Simpson, F.R.S., Director of the Meteorological Office, who had been one of my grandfather's mathematics students at Manchester University and had subsequently been south to the Antarctic with Scott's expedition. But when August came, and I had heard nothing more, and had had no replies to any of the other job applications I had made, I no longer dreamt that any of them would be successful and went away on holiday. Having asked my father for a few pounds - I believe he gave me £17 - to cover the costs, I re-joined the Scottish Youth Hostels Association. I determined to make the money last for as long as possible. On 5th August 1936 I sailed with the Dundee Shipping Company vessel from near Tower Bridge on the Thames to Dundee for 36 shillings. My meals cost another 11 shillings.

On arrival early on the 7th, I took the ferry across the firth to Newport on Tay and set out on foot with my rucksack along the south shore, heading inland. Some way on I got a lift in a truck from two men who had been "lifting tatties" (digging potatoes). They generously shared a few plums with me. When they reached their destination, I walked on through an attractive wood at the waterside, but soon had to give that up when the rocky ground and undergrowth became too difficult. So I continued on for seventeen miles along the hilly road to Abernethy, where Meg Elliott and her sister and family, the Patons, were expecting me. Their tiny daughter, Janet, was too shy to make friends until my third day with them. Her older cousins, Jean Elliott and brother were there too and keen to hear of my adventures so far. Next morning we all walked up the hill behind the village with great views north across the Firth of Tay into the Highlands. On the Sunday (August 9th) we went up the Pictish watch-tower beside their garden wall. A few days later, Jimmy Paton, and Jock Elliott and a friend, Alec Buchan, whose grandfather had been a famous meteorologist, took me with them by car, heading north, to a permanent camp they had in Glen Feshie. This was my introduction to Speyside, and I put up my little tent beside their camp. Three other friends were already installed in the caravan and we spent the evening chatting in the farm kitchen at Tolvah alongside. Next day, we crossed the Feshie river by the swinging wire bridge and climbed the main hills on the east side, then along the ridge to Sgoran Dubh (3,635 ft above sea level /about 1108 m) with its fine view of the high Cairngorm summits, Braeriach and Cairn Toul (both well over 4,200 ft). We walked on along the ridge overlooking Glen Einich with its lovely loch and so down into Rothiemurchus forest, one of the chief remnants of the primeval Caledonian pine forest. We continued through the wild tract, with many old juniper trees and heather as well as pines, to round the beautiful Loch an Eilein with its ruined castle on the island in the lake. Farther on, we were picked up by car and driven back to the camp, after six hours on the hills and in the forest. The next day, after a night of heavy rain, we were driven around to Coylumbridge, near Aviemore, and walked up the path beside the river Druie, which gets gradually rougher and finally leads out onto the mountain over rocks and boulders to the Lairig Ghru, the main pass through the Cairngorms. This impressive defile cuts straight through between high precipices below the highest tops. One party that died here on a winter trek in the nineteen-thirties had a minimum thermometer with them that was registering -29°C when the victims were found.

    A few days later I said goodbye to my friends in Glen Feshie and, with my rucksack and the folded tent on my back, I set off - alone once more - up the glen. An eagle was circling over one of the tops near by. This time I went right to the head of the glen and on into open lonely country - an endless moor of peat and rocks and heather - to follow the long course of the Geldie Burn down to Deeside, a tramp of over 25 miles (40 km) that finally led me to the motor road at Linn of Dee.

    Word from my friends in Oslo, whom I hoped to re-visit, had reported a Norwegian student's success earlier this summer in hitch-hiking three or more thousand miles around northern and central Europe, including a visit to Britain, and it struck me that it should be very interesting to try for myself the same method of getting along. These were pioneering days for hitch-hiking, but there were possibilities in 1936 that would unfortunately be best avoided as unjustifiably risky fifty and sixty years later. As it turned out, my holiday trek that summer covered about 2,500 miles by sea and over 1,000 miles on land at a total cost to me of about £16. Obviously, this owed a deal to the hospitality of old friends and the kindness of many others along my way. There was fascination in the wide variety of folk I met and the many ways of life I saw something of. And nobody was persuaded to go out of their way or do anything they did not want to do.

Getting lifts that day in Deeside was no problem. I was soon taken to Braemar, the first village, where I got down and had a welcome tea in a tea-room. And with the next lift in a then very up-to-date Chevrolet car, putting me down in Ballater, where I slept in the youth hostel, the driver thanked me for my company. One more lift, the next day, saw me to Aberdeen. To find a night's lodging, I went to a Mrs.K??????, recommended by the youth hostel man in Ballater. The door was answered by the man of the house who introduced himself as Lt.Col. K?????? (retired), recruiting officer. The rooms were decorated with military portraits and battle scenes. It turned out to be a house of corruption, every bed occupied, even in the afternoon, by unmarried couples indulging in sexual activity. Next morning a boy of about 6 or 7 in the house told me he wished he was going to sleep with me. So I decided it was time to leave. I paid the few shillings asked for my night's lodging and was encouraged to "come back next year and bring your girlfriend".

I went round the harbour in Aberdeen and found no boats bound for Norway. I was told by the dockmaster in Aberdeen that the chances would be better in Buchan's salted herring ports, Fraserburgh and Peterhead, on the coast of north-east Aberdeenshire. It was just the height of their season. So I went on there. Mrs. K?????? generously gave me tea with meat sandwiches as a send-off, so kindness still had a recognized place, even in that household. Two lifts got me over the 30 miles to Peterhead, and the second one found me lodging with Mr. & Mrs. Frazer, fruiterers, who gave me a good supper of fresh plaice hot, and fruit, scones and jelly. I spent all the next week wandering round the little ports along that coast as far as Banff, meeting many fisher folk, boat owners and brokers, learning about the trade, how the herring was shipped to the Baltic, to Stettin, Danzig, Memel and on to East Prussia and Estonia. There were seafaring tales from as far as British Guiana. One night I "slept" on the stone floor of somebody's wash-house, but I got good smoked fish for my breakfast and the lady refused any payment. On the Thursday (20th August) I got the first word of a boat coming from Hamburg, coaling at Methil on the Firth of Forth, intending to pick up a cargo in Fraserburgh for Norway at the weekend. Despite the frustrating delay, and occasional upsets through constipation and feeling sick, it was an interesting time and a happy one because of the interest shown by these folk in my experiences and plans - I had my photograph taken several times - and because of continual amazement at their generosity. I learnt that "bread" hereabouts means "oatcakes" and what is known in English as "bread" is "loaf". The boat in the end did not leave until the Monday and I spent the last couple of nights at the picturesque little port of Pennan and did the 20-odd miles to Fraserburgh, sailing in the evening at 7 p.m. I was accepted as a passenger upon payment of £1 to the mate, at his suggestion, for the use of his cabin, which he readily gave up to me. The Norwegian vessel I was now on was really a 90-ton trawler serving as a tramp ship, carrying a load of barrels of rotten fish to be converted into herring manure. The barrels filled the deck right outside the cabin door. There was a considerable gale that night, the sea became very rough, and with the rolling of the little ship, two or three of the drums of rotten herring were lost overboard. Prudently, the mate had instructed me not to go out on deck at all until we reached sheltered waters. He showed me the wash-hand basin in his cabin and told me not to use any other toilet. "Da man er tilsjøs, må man skifte sig som best han kann"(when one is at sea one must make shift as best one can). He was so good as to bring my meals to me. There were interesting discussions about food and also about language. The Norwegians were shocked, as I was, by the impossibility of buying any better bread in that part of Scotland than Beatties' Bread (distributed in those days daily all over Scotland in its paper wrapping, not particularly palatable to those who like crusty bread). At the little town where we arrived in Norway, on the island of Stord, off the west coast, they were used to splendid crusty white bread from the local electric bakery. I asked the Norwegian crew how they found the Scots language as generally spoken in Buchan, many of the words being close to Norwegian forms, but they did not notice this, since the pronunciation is very different. They would only say: "This is not the English we learnt in school".

    The same trawler, having unloaded in Stord, took me on to Bergen. While in Bergen, I stayed in the youth hostel which, as in other towns at that time, was in the attic of a big hotel. At a meal I happened to sit down next to an official of the Norwegian Youth Hostels Association (NUH), who explained the association's policy to me: "We believe that if you come now when you are young and haven't much money, you will come back again later in your life, when you have made some money, and pay the full price." And so it turned out in our family, except that we made so many friends in

Norway and elsewhere in Scandinavia that we have also enjoyed much private hospitality and returned it when we had opportunity. In Bergen, the trawler skipper introduced me on arrival to Leiv Olsen, a ship's broker living in the town. He quickly secured for me a nearly free passage as a supernumerary crew member on a cargo ship, the Føyen, 2,400 tons, bound for Trondheim. We followed the mostly well sheltered shipping lane, northbound up the coast, which passes nearly all the way between islands, many of which are hilly or mountainous, and the generally mountainous mainland. The scenery is very fine, especially near the high cliff of Hornelen on one of the bigger islands. This cliff almost overhangs the shipping route at a point where it is narrow and takes a sharp, nearly 90° turn. Ships have been forbidden to blow their sirens at that point for fear of dislodging rock. We had sailed at 10 a.m. on a dull morning, the 29th August, with all the hills clouded. The captain apologised to me that I would have to put up with sjømannskost (seaman's food), but in fact it was the tastiest food I had had for a long while: anchovies in a mild fluid, herrings and other pickled fish delicacies in little tins that have since acquired an international popularity.

The captain told me, having invited me to eat at his table, that his firm was particularly good for taking people on nearly free passages, though they were not allowed on trans-Atlantic runs. In the afternoon, he and I sat together doing a Norwegian crossword puzzle in the weekly magazine *Hjemmet* (the Home) and completed about three quarters of it.

On the second morning out from Bergen, Sunday (the 30th), I woke at 7 o'clock when the ship's engine stopped, and I found the captain in a fury over the pilot who could not answer his first question as to exactly where we were in the coastal channel, a very dangerous situation to be in because of the number of submerged rocks. On the captain's orders we then went slowly astern for one and a half hours, following as exactly as possible the way we had come. Only then was our position fixed, making it safe to resume our course northward. Back then to a breakfast of ham and eggs, coffee, anchovy paste, liver sausage etc. After breakfast I went back on to the bridge. Despite the drizzly weather, a good view of a school of porpoises. No let up in the thick weather, even when the wind veered north-west. The captain, still furious with the pilot, sent him down to his cabin to sleep, even though it is illegal to sail this coast without a pilot on duty.

We reached Thamshavn that evening, where the ship is to unload sulphur from Finland. Already 30 km farther west, the forest was noticeably damaged by the sulphurous air in the fjord, which has led to litigation. Next morning I went ashore with the local ships' broker and arranged by telephone for the local bus to stop and take me to Trondheim (35 km, fare

2kr.20). So I had travelled from Bergen to Trondheim for 6kr., including two days food and lodging plus schnapps, everything first class!

In Trondheim I stayed in the NUH youth hostel and explored the city, admiring the streets lined by handsome wooden terraces, two storeys high, in the dignified and well proportioned pan-European style of AD1680. (As in some other places, in other European countries, these harmonious streets had been built complete in a few years after a great fire had destroyed the earlier buildings - a positive benefit of a fire in such a good architectural period! I also viewed the fine Romanesque cathedral, for centuries the pride of the North, the great head church of medieval Christendom in northern Europe, a place of pilgrimage from far and wide. A tragic result in 1349 was that the pilgrims from central Europe and beyond spread the great pestilence, the Black Death, along their way, for example, through central Sweden, whereas there were many places deep in the vastnesses of the Norwegian mountains, off the routes to anywhere, to which the plague never penetrated.

In the afternoon of Tuesday, 1st September 1936 I left Trondheim for Oslo, first taking a tram to the city's edge. Hitch-hiking at first proved difficult. No lifts at all in the first ten kilometres, and the next two lifts were of just 1 km each, the first one in a horse-drawn vehicle. And just 32 km farther along the road, being very tired I turned in at a farm where they let me sleep in the hay loft above the stable, and I got a breakfast of egg, coffee, and barley porridge, which the family had begun eating cold the previous evening. The farmer had been twelve years in the USA before coming back to Norway. A few kilometres farther on, a schoolmaster and his wife flitting in their old Chevrolet van, with all their belongings, to south Norway in hope of finding a better paid job, took me 50 km on, through Støren and Oppdal into the hills. We were leaving the grain harvest fields behind, but coming to much finer weather. I had a main meal with them at a cost of 1kr.65 and said goodbye. I walked on along the road through the forest and past a few farms. After 7 or 8 kilometres, I was picked up by a young geologist driving a comfortable, fast American car and had difficulty in persuading him not to take me all the way to Oslo (a distance similar to that from Edinburgh to London) that very evening. He had driven the whole way a dozen times that summer and wanted company! But, for me, there was far too much fine country which was new to me, that I wanted time to see along the way. So I thanked him and got him to put me down near Hjørkinn station on the Trondheim-Oslo line on the high ground of Dovrefjell, a wild open moor with the white reindeer moss among the other low plants.

There were occasional saeter huts and here and there a few odd cows grazing on the moor. I wandered over to one of these huts, more of a cottage than a hut, the highest up one of those on the north side of the watershed. It was red painted and obviously newly built, with bare light wood on the walls inside. I was keen to sample the old saeter life, the summer grazing of some cattle on the heights, and here was my chance. I found just two girls of maybe slightly more than my own age, sisters, in charge, spending the summer up here, nearly 1,200 metres (4,000 feet) above sea level from their family farm in the valley to this summer pasture for the twelve cows they had brought with them. With their permission I rolled out my sleeping bag on the floor of their hut. I was struck by the generosity of these saeter girls and their confidence with me. They served me with their own simple food, the rømme (thick cream) porridge and fruits of the moor, politely saying "Veirs'go" (please) and "Berre spis" (just eat), while the full moon shone in the clear sky outside on the white (snow-capped) head of Snøhetta (2,286 metres or about 7,500 feet above sea level), the highest point in this wilderness view. We talked a little in Norwegian about life here and in the big cities, London and Oslo, which was the farthest they had been. Next morning I was up at 6 a.m. for a breakfast of milk and cakes and coffee, while morning mists played around the juniper bushes and across the more distant views. I climbed to a couple of high points to get the best views across the moor and the Hedmark hills, but I was too late to see the cows go out to their grazing. And I was deeply embarrassed when the girls absolutely would not have any payment for my stay.

The saeter life has dwindled since, and been largely abandoned. In some areas, particularly in the nineteen sixties, very occasionally an odd bear or a wolf has re-appeared after a gap of fifty years or more, where the huts have been deserted. In other parts, the huts have been made over into holiday homes for winter sports and summer walking or fishing trips.

The day was so brilliantly sunny and really too hot for a long walk. So I got more lifts, the first one as far as the major road junction at Dombås. The second one took me on down into the great central valley of Norway, Gudbrandsdalen, at first through pine and fir forest and later, south of Ringebu with its old church and steeple, passing many great farms, their big, prosperous looking farmhouses well up the hills on the sunny, south-west facing side. I had soon come past Dovre, where I had thought to spend the night. Once again, the full moon shone over the trees. Dovre is where the descendants of the old king Harald Hårfager of a thousand years ago have lived ever since. And some way farther on, near the valley junction at Otta, was the farm owned in the fourteenth and fifteenth centuries by Lavrans and Kristin in Sigrid Undset's wonderfully perceptive story. Looking back over sixty years to my experiences here in the nineteen-thirties, I am struck by how well I was received alike up on the saeter and in the main farms in the valley. There were always folk then who had time to spend talking with me, at least in the evenings. At Ringebu I was shown round the twelfth century stave church by an old woman caretaker speaking all in a dialect which I could not understand much of, but it was interesting to see its small windows and the many crucifixes from Catholic times as well as a book, written by a seventeenth century (Lutheran) priest in an old Gothic handwriting, describing King Frederik IV's crowning. At Ringebu I had some difficulty finding a night's lodging, as it was already getting dark when I arrived. At the first two farms I called at to ask if they had anywhere I could roll out my sleeping bag, the people were too frightened by having a caller, however polite. I blamed myself for having left the matter so late that I worried the people I asked. However, at the third farm I tried, there were just three women and some children who were good enough to take me in. One of the wee girls at her bedtime said to me prettily "God natt da på deg" (Good night upon thee). I slept in the hay in the big barn, and I was up next morning about 8.15 a.m. and was given breakfast with porridge and coffee and was charged just 50 øre (half of one krone) for my stay.

When I left Ringebu, I was hoping for a lift to Lillehammer, where there is a very fine collection of ancient buildings from the Middle Ages and after, mostly built of wood, the earliest ones very small and with just the earth for a floor, in the open-air museum among the trees at Maihaugen. There is also an eleventh century stave church in the collection and some later - seventeenth to nineteenth century - two-storey farmhouses. It is a very colourful museum to see also because of all the traditional costumes, many of them with much of the bright red of the Norwegian flag, but also much white lace and blouses to be seen and much of the silver workmanship of the medieval smiths and jewellers. But I was fated to have to leave all that to visits in later years, because my first lift from Ringebu, with a family in a saloon car, took me all the way to Oslo (253 km or about 160 English miles). Their conversation already displayed their anxiety about the coming World War and the problem of "the troubles brewed up by the nations farther south", a recurrent theme in Scandinavia. They were so kind as to drive me to the door of the Norwegian Youth Hostel (NUH) in the Missionshotell. But it was full! So I telephoned the only person I then knew who lived in Oslo, a girl called Tunny Müller, who had spent some months staying with my friends, the Kellocks, who lived in the Chiltern Hills near Missenden. John Kellock worked in Hambros Bank in the City of London, and had frequent business connections with Norway and loved the country. He had also stood as a Liberal candidate in the last general election. His wife was a member of my Norwegian language class at University College.

Towards a Dead End

Tunny had only returned home earlier that summer, and I had arranged to call on her in Oslo. When I telephoned, she and her parents promptly invited me to bring my rucksack and sleeping bag and stay in their flat overlooking Frognerpark in the west end of the city. She was one of a large family, and I walked around many pleasant parts of the city with her, her father, a district doctor lately retired, and one of her brothers or sisters, getting to know the place well. One Sunday, a party of us including two of Tunny's sisters went on an all day hike through the forested wonderland, Nordmarka, that stretches far out over the hills just north of the city, past lakes in the woods to a viewpoint, in Krokskogen, near the farther limit of the area to the north-west where the land drops to a huge lake, Tyrifjorden, with many arms.

My long friendship with the Müllers' big family introduced me to a fair range of the strands of Norwegian life, some of them already typified in the previous century in the plays of Henrik Ibsen. Their mother was a neat and competent little lady from More county, north-west on the coast towards Trondheim. Tunny's eldest brother was the first mate on a liner "m/v Meteor", sailing the seas of the world who paid a visit home while I was there. Another brother, Karl Wilhelm, became the pharmacist in Haugesund, a port on the west coast south of Bergen. A slightly older sister, Dorothea, was already secretary of the Norwegian Tourist Forening (Union), dealing with facilities for walking and skiing in the countryside, while Tunny's close friend, Fia Dedekam, whom I also met in England, the daughter of a big ship owner on the south coast near Arendal, was to become the wife of a cabinet minister in a Norwegian Labour government. Fia and Tunny were skiing companions around Oslo until the nineteen-eighties.

During the war in 1940-45 the eldest brother escaped, at his second attempt, to Shetland to join the free Norwegian Navy, while his less fortunate sister, Dorothea, was taken by the Gestapo to Grini, the concentration camp outside Oslo. But, at the time of my visit in 1936, people's most immediate worries were about completing professional training and finding suitable employment once qualified, a problem that was particularly acute for those training in medicine and dentistry. One result was the phenomenon of the *"ewiger student"*, the everlasting student, working more and more years to get more and more qualifications, but never getting a real job. My friends were also worried about the breakdown of morals, particularly the breakdown of the sense of obligation to friends, relations and wives, that in their view was tearing away the heart of society. However, it was while I was still with them in their tiny flat in Oslo that the solution to my own most pressing problem arrived. In the middle of September a postcard arrived for me from my father telling me that the

Meteorological Office was offering me the job they had interviewed me for. I wrote at once to accept. But finding a passage on a cargo ship at short notice was not easy. One firm explained to me that the summer herring fishing season in Scottish waters was now finished, so that only an occasional Norwegian ship would be going over. Instead of waiting, I got a place on deck to roll out my sleeping bag on a ship bound for Copenhagen. Two of the Müller sisters, Kari and Dorothea, came to the harbour to see me off on the afternoon of 15th September, as Tunny was still at work. And the whole Müller family, when saying goodbye at the flat, thanked me profusely for coming to see them - so profusely that I could hardly thank them properly - and they added, pressingly, that I must come back to see them again. I went on by train and boat across Denmark and the North Sea as a normal fare-paying passenger, and by the evening of 19th September I was at home with my parents. My father was unwell, and I believe this was the start of his nervous breakdown which kept him at home off work, and sometimes in bed, for a number of weeks.

    I have to wonder what part, if any, the difficulty of my relations with him may have played. No such thing was, in fact, ever suggested to me. But it has seemed to me likely that the intensity of his ambitions for me, and their all too specific nature, amounting to a set determination to get me to follow a particular line of study and to acquire specific expertise that I did not feel able for, led him at this stage to a feeling of failure that he could not understand nor accept. But my mother was very well and excited about the success of my holiday trek. The trip had certainly done wonders for me in restoring my self confidence

CHAPTER 7. BECOMING A METEOROLOGIST

So I was to be in paid employment at last. The salary of £280 a year was typical enough at that time for new entrants to the Civil Service with scientific degrees. (The value of money was very different from what it has since become, nearly sixty years later, and I used to cash a cheque for just £2 each week for living expenses when I was established in lodgings away from home.) And on 12th October 1936, I joined the national weather service, the Meteorological Office, at the old Croydon Airport, which was then the "hub" of the international airlines coming into London, but has since been swallowed up by the growth of south London and has now long been buried under bricks and mortar. I was one of the second batch of six university graduates taken on in that year on account of the intended expansion of the Royal Air Force. It was a time of severe unemployment, and we all owed our jobs to the government's re-armament programme. Our instructor, a slightly built, very earnest man in his thirties, was Mr. S.P. Peters, who had worked for the airships development, which was abandoned after the disaster to the R101 over northern France in 1930. He was a very lucid, patient lecturer and a very kind, conscientious man. His precepts were also very firm and practical, insisting on the use of clear, precise language in wording information and forecasts, banning the various sorts of slipshod, subjective language - such as "reasonable temperatures" - often heard in radio and television weather forecasts in the nineteen nineties. There is also clearly no justification for another favourite expression of our latter-day forecasters, "would you believe it?", for example after speaking of an expected snowfall in March or a frost in May or June, which has probably occurred ten or more times in most parts of Britain in any fifty-year period one might examine. Peters' lectures on weather analysis depended heavily - as was nearly universal in the nineteen-thirties - on the Norwegian discovery in 1917, due to Professor Vilhelm Bjerknes, and his small team of co-workers in Bergen, of the fronts between windstreams of unlike origins converging in the atmosphere and leading to recognizable, and for the first time physically explained, patterns of development of weather systems - depressions, cyclonic storms and anticyclones. We were also introduced to the practical side of making weather observations and reading the instruments, including of course the use of the theodolite to follow a hydrogen-filled "pilot balloon" to track the upper winds. Beyond these tasks we learnt how to draw and analyse weather maps from the collected observations that come into a meteorological office. And large slices of our time were devoted to studying the lines of equal barometric pressure (isobars) on weather maps covering the whole width of the Atlantic Ocean day by day in many previous years. The direction of the lines gives a picture of the wind direction and their spacing enables one to gauge the wind speed, since the gradient of pressure controls the strength of the wind. These relationships work best over the open sea or in the free air at greater heights (but for that one needs to map the winds and pressures at greater heights). The object of our work was part of the preparation for opening the first commercial passenger air services across the Atlantic, which in fact was achieved in July 1939. From our measurements of the barometric pressure gradients we were able to tabulate the likely times needed, day by day, by the aircraft of that period to cross the ocean between Europe and North America in either direction. From this tabulation the routes that would be feasible at different times of the year could be decided and likely flight schedules envisaged.

When the Trans-Atlantic air service was inaugurated, using flying boats of the Sunderland and Yankee Clipper classes, the normal summer route was the direct, short crossing from Foynes, Co. Limerick, on the lower Shannon in Ireland, where the river was two miles wide and there was an anchorage effectively sheltered by Foynes Island, to Botwood harbour on the north coast of Newfoundland. The distance was about two thousand miles. Over the rest of the year, when head-winds on the direct route were sure to be prohibitive on some days, more southerly routes were used, generally going to America from Foynes by way of Lisbon, Bathurst in North Africa, and Bermuda to New York, and returning via Bermuda, the Azores and Lisbon. For each of the scheduled stops there had to be alternatives available within about two hours flying time in case of bad weather at the intended destination. The main "legs" of the southernmost route usually offered following winds (the Trade Winds) for America-bound flights, especially in the winter-time, whereas the direct route was in the latitude zone of prevailing westerly winds, meaning that mostly head-winds had to be expected on west-bound flights for most of the year.

At 5 p.m. each afternoon, when our working day at Croydon Airport ended, I walked back over the low hill from the airport by the footpath beside the old Croydon bypass road to Purley, where my lodgings were. Just before reaching them I had to pass a little, unpretentious Austrian cafe, and every day I went in there for a cup of their good coffee, with thin coffee-cream, and a rum-baba cake: quite a luxury but an inexpensive one!

As that winter of 1936-37 neared its end I had a problem: how best to use the last five days of my 1936 annual leave entitlement, which had to be taken before the end of February or lost. So I decided to take advantage of the fact that our training school was at the London airport of those days. I went to each of the airline companies operating there to find out which one would offer me the best discount on their air fares. The clear winner was Deutsche Lufthansa, the German airline, which proposed to charge me nothing but the £1 each way required as insurance money for flying me to and from Berlin (where I had never been). So in mid February 1937 I took off in a Junkers 88 plane, seen off by my parents. We put down at Amsterdam's Schiphol Airport, which was not then the imposing place which it has since become: in that winter it was just a very big, flat, and particularly wet, muddy field. When the aircraft revved its propellers to take off again, the violent airstream they generated set up great clods of sloppy mud that plastered the plane like so many cow pats. We then flew on, keeping very low, only a few hundred feet above the ground, below the low clouds that constituted a complete overcast. In Holland one could see everything that was going on on the ground below. I remember an old farm worker with a yoke on his shoulders, carrying a pail on either side, across a muddy farmyard to some animals. And over the plain in North Germany one looked across to the Weser hills with their partly cloud-covered tops more or less level with us. Little did I think that some fifteen years later I would be living for some time in Bückeburg at the foot of those hills. When we reached Berlin, the crew of the aircraft kindly recommended me to an inexpensive hotel beside the River Spree. I put on my rucksack and set off to reach it by tram and on foot. On the way, the streets were swarming with soldiery, mostly in army (Reichswehr) uniform and mostly very young. On my way into the centre of the unexplored city I was stopped by a young soldier in uniform, who saluted me and asked me (in German, of course) if I could tell him the way to the Brandenburg Gate, which was later to become well known as one of the check points on the boundary between the eastern and western sectors in the very centre of post-war occupied Berlin. So my questioner revealed that he was as much of a stranger as I was!

The weather continued overcast, muggy, occasionally drizzly, and very mild (over +10°C) throughout my visit to Berlin, but the trams were still secured against cold draughts by heavy felt hangings over the bottom edges of the windows, and the tram drivers still wore their furry leather gloves up to their elbows. I could not keep my bedroom window in the hotel open: every time I went out at all someone shut the window. Berlin was plainly organized on the basis that winter must be very cold (as had been more often the case in the previous century). There was a somewhat disturbing incident at breakfast in the hotel, repeated each morning, when a middle-aged German, looking like a business man but with his face marked as so often in earlier times by duelling, joined two colleagues, greeting them with "Guten Morgen. *Gott strafe England*" (Good morning. May God punish England): but I reflected that this ritual really mirrored that of my father's war-time friends at dinner parties in our house only a year or two earlier with their unrelenting repetition of yarns about the dreadful "Hun". How could mankind ever achieve an era of peace while such attitudes were kept alive? (But surely there is still in the nineteen-nineties in Britain a continual re-run of material on radio and television, and in books, re-living the tales and experiences and attitudes of the war years - a seemingly insatiable appetite for war reminiscences and violence!)

CHAPTER 8. AT HOME IN SCOTLAND

Soon after that, in March, at the end of my six months training, I was posted by the Meteorological Office to a Royal Air Force flying training school airfield on the east coast of Scotland at Montrose. There were two separate duties that I had there. One was to forecast the weather a few hours ahead for the flights of the young RAF pilots under training. The other was to study the very awkward problem of the sudden fogs, or *haars*, that come in from the sea on that coast. There was, however, a serious hitch at the very beginning of my time there. I went on a preliminary visit to Montrose, travelling north from another RAF airfield, at Leuchars in Fife, where I was to spend six weeks familiarization with a much more seasoned forecaster while living in St. Andrews. There were a dozen or more wooden pegs in the grass on the front of the airfield at Montrose marking where the hut that was to house the Met. Office would be built, among the other offices in the whole encampment of huts, and I had brief introductory chats with one or two of the officers and department heads. But when I came back just before the day in early May 1937, when the Minister for Air announced in parliament that six new Met. Offices, of which mine was one, had been opened on RAF airfields, all that had in reality happened on the site was that the wooden pegs had been removed and the waving grass was as undisturbed as it had ever been. So my assistant, Angus Young, and I were allocated an unoccupied hut (which had been furnished as a dormitory, with about twenty beds) and the heavy, antiquated typewriter, which was the only other item that had arrived, was placed in solitary grandeur on one of the beds. It was a bit of a responsibility to decide how best to employ ourselves in connection with the tasks ahead and, in particular, the fogs investigation. That was also an awkward question to answer a good many weeks later when I was asked by an interview board called to judge whether I could be established on the permanent staff of the Meteorological Office how I employed the assistant in those circumstances.

What we in fact did with our time was to tour the countryside of Angus and Kincardineshire on our bicycles, between the sea and the main mass of the Grampian hills as far as the heads of the nearest glens, visiting all the local schools of which a real network still existed at that time, and persuaded the teachers to get their classes to keep records of the dates and times when fog appeared and when it went again. On one of our biggest days, we covered well over fifty miles, including crossing one of the main hill ridges, carrying the bikes up a vague path through the heather but free-wheeling on animal tracks through the shorter heather down the other side.

On the uppermost part of the climb there was no alternative to carrying the bicycles because the heather was so long, although the path was marked on the map. Making our way home to Montrose, we used the old Cairn o'Mount road over the hill, its summit at 1,488 feet (453 metres) above sea level commanding a wonderful view in every direction of the hills and lower ground and coast. It is a fine wild road over the heights, unfenced but marked by high posts for occasions of deep snow, and in those days it was an unmetalled road. I once saw a big capercailzie roosting, apparently all unconcerned, on an upper branch of one of the nearest pines to the road. At the last school on the north side of the hill, in Glen o'Dye at the forest edge, where we called to arrange the described recording of fog occurrences, the kind schoolmistress, a lady of maybe forty years, made us a splendid Scottish tea of scones and baps and jam and we all sat together in wide-ranging conversation for a long while and so set us on our way in good heart. Finally, in the light May evening, we went on over the top and down to Hillside and Montrose. I felt concerned in the weeks that followed, when Angus Young developed water on the lung, in case I had led him to overstretch himself, but I suspect that his cigarette addiction may have had more to do with the cause than our big day of possibly excessive physical exertion. By the following autumn, Angus had made a good enough recovery for us to try ourselves out as shooters with borrowed two-barrel guns and go back to the Cairn o'Mount to shoot some of the wild Arctic hares that turn white in the winter-time. We chose a foggy day with the cloud well down on the hills to hide our misdemeanour. But we did the hares no harm - for which I am glad - as we could not see the hares well enough to aim successfully, and, when for once we got near enough to a hare sitting up and sensing danger, we were so surprised that we did not think of the guns in time and never fired at all. It was our only shooting exploit.

  Over the next weeks our office building went up, the furniture and the instruments and blank weather maps that we needed to do the regular weather observations and forecasting job duly arrived. And we began to get weather reports by radio and teleprinter to make several weather maps each day, covering the British Isles, much of Europe and the Atlantic Ocean as far as Iceland. Those maps - although they were the standard supply at that time - were not really adequate, because when new weather systems came on the scene from the north, near Iceland, they could reach us in eastern Scotland within twelve hours or so.

  It seemed certain, almost from the outset of our work, that the sudden onset of the haars or sea fogs of that coast depended on the development of the day-time sea breeze. The circumstances over some wide region had to be suitable for that development to take place, including the moisture of the air over the land and the sea, the gradient of temperature with height, and (perhaps most importantly) whether there was too strong a flow of the general wind across the region to allow such a locally generated development as the sea breeze to occur. It was going to be important to get measurements of the temperatures at different heights above the ground and above the sea surface, since they largely determine the density of the air and so indicate whether any air beginning to rise over the warming ground will be encouraged to go on rising or not. So I had a good many flights over the surrounding areas with one of the flight instructors. But he and his colleagues were also young and keen on flying for the fun of it. So it was always difficult (or nearly impossible) to get them to keep the plane flying at each level for long enough for the thermometers to settle to a steady reading. It was so much more fun to take me up to 15,000 feet, then turn the aircraft on its side and let it go into a free fall of some thousands of feet before zooming away to somewhere else. The view from the open cockpit of the aircraft was always exciting. But we did amass enough useful measurements for me to be able to write a report, entitled "*Haars, or North Sea fogs, on the coast of Great Britain*", which was well received by the authorities. Unfortunately, it was soon decided that, because of the value of the report for forecasting for military flights, the report should be immediately classified as 'Secret'. This was a somewhat disappointing outcome from a year and a half's work for a young scientist keen on establishing his worth, since the more widely known any piece of research becomes, the more it must help one's career.

    One day the flying authorities and I were confronted with a mystery that caused much immediate anxiety, and the commanding officer of the airfield at once called me in to advise him. It was towards 10.30 a.m. on a bright day in early summer, when all the trainee pilots had just gone off on their morning exercises. The sky quickly began to look ugly over the sea to the south-east, and the air became murky right down to the sea surface. I was asked the obvious questions: "Is it fog?" and how soon would it be blocking all the view. I had not mentioned fog in the day's forecast, and I at once answered: "I don't see how it can be fog as the air is quite dry, only 60% of saturation. But can I go up in one of your 'planes and have a look at it?" Soon, I was airborne, and by the time we reached a height of just a few thousand feet the solution to the puzzle was obvious. From that height, the air which is habitually very clear over eastern Scotland (visibilities of fifty to a hundred miles, and more, being commonplace) revealed a great smoke trail, starting up as a low, wedge-shaped gathering of pollution from the myriads of domestic house-fires and industrial chimneys in the densely populated belt in the region of Glasgow, Edinburgh and other lowland towns, and very gradually reaching up to greater height the farther it proceeded. It was being carried by a general west wind through the Clyde-Forth valley out on to the North Sea, where at some distance out it began to be caught into the developing sea breezes of the east coast. In that way it was turned around and approached the coast between Montrose and Aberdeen from the south-east. Soon, of course, this dry haze arrived and enveloped the coastline and the Angus and Kincardineshire countryside. But one could still see one to two miles when it was thickest.

It became abundantly clear after a few weeks observation that in all normal cases involving the arrival of a real sea fog, after clear, sunny weather during the earlier morning hours, the sudden onset of fog overwhelming the coastland was linked to the arrival of the daytime sea breeze. The timing was essentially the same as the arrival of the sea breeze on days that remained clear and bright.

Despite misgivings over what would come of my work being so closely related to military matters, my time based at Montrose was one of the happiest periods of my life. On arrival in Montrose, I stayed the first few nights in the oldest, but attractive-looking stone-built Corner House Hotel next to the Old High Kirk with its tall, elegant steeple and warm-coloured old red sandstone at the top of the High Street in the centre of the town.

It is a fine town with its broad sloping High Street and two other streets radiating from the steeple church which dominates the view because of its height. There is also a good harbour, which was at that time somewhat neglected, but has since been rejuvenated by the North Sea oil industry and the vessels that serve it. There is also a two miles broad tidal basin behind the town, so that the main coast road to Aberdeen and the railway come in over bridges over the River South Esk between the basin and the harbour. The kind, motherly lady, Miss Mitchell, who then kept the Corner House introduced me to her friends, George and Effie Gray, who had just retired from their farm at Boddin Point on the coast just south of the town and were looking for a lodger in their new home in Hillside village, a little to the north-west of Montrose. I went to live with them for two very happy years. Their stone-built house, with its cheerful white-framed windows and well-stocked garden, had a magnificent view, down across the flat land and Montrose basin, to the higher ground and trees beyond and to Scurdiness lighthouse at the point beyond the harbour mouth. It was a view, too, that at that time was full of activity with two mainline railways converging just out of sight on their way to Aberdeen, the one from Edinburgh and the other from Glasgow, with a mail-bag pick-up point right below the house. There was also a little branch line that connected with the Edinburgh line near Montrose station and came in a great curve across the flatland to join the

Glasgow line at a little station half-a-mile away, Dubton, just below our village. This whole railway network was in sight beneath our windows. The continual steam trains could be watched at any time of the day, and their white smoke in the sunshine showed clearly what was happening to the winds. I once watched the formation of a depression from our sitting-room window, shown by these smokes. It was a day when there had been some hours of rain until a slow-moving front from the north very slowly cleared and a little bright sky came in view from the north. Soon, however, it became noticeable that the cloud sheet was not making any further progress south, and the smokes from the trains, which had shown a westerly wind, showed uncertain drifts and then began to reveal a persisting drift from the east. There had been no east wind near Scotland on the weather maps, but there it was starting before our eyes. Next, the sky to the south of us became greyer and some lower, scud clouds appeared and soon it was raining again.

Besides Mr. and Mrs. Gray and me, our household included their one son, Sandy, who was about my own age. He toured the countryside up to 40 or 50 miles around in his car as a seed salesman for Kinnaber Mill, just two miles down the North Esk river, the other side of the village and its hill, towards the sea. He became a good friend of mine and some of his friends too, but they all died when they were no more than about seventy, perhaps because of excessive cigarette smoking. We went to the farmers' balls together, many of the dances being the traditional Scottish dances, and there was always a cup of hot soup at the end to help us on our way home at 2 or 3 a.m. Sandy and I also played golf occasionally, but both decided to give it up on the same day, after a disastrous round, which made us realize we preferred to have the walk without the wretched ball. It was a beautiful nearly cloudless day. At the start our mood had been sunny and carefree too, but shamefully missed shots and all the frustrations of the game had spoilt all that.

Sandy's daily rounds with his car frequently gave some cause for amusement, as when he came home with a lobster which he had bought from a fisherman at the harbour in Gourdon, north up the Kincardineshire coast. He had paid one shilling and sixpence for it, or 7½p in today's money, and was troubled lest he had paid too much. He was a good, responsible driver and had a shock one Saturday evening when driving quite slowly, in traffic, into Montrose. When he was forced to stop altogether for a minute, a cyclist's hands came slowly sliding down his windscreen from the top. The dropped-handlebars rider had been close to the tail of Sandy's car, which had a smoothly rounded body profile in the fashion of 1937. With no sharp edges to catch his slide, the cyclist was unharmed but whether he learnt anything about riding too close behind cars we never knew.

At Home in Scotland

That winter, while on a short visit to the Patons in Edinburgh, I bought my first car, a Morris Cowley coupé, of 1930 vintage, heavily built and very strong, as was usual then. It cost me £7 at a garage in Corstorphine, and I was its fifth owner. It was known at the RAF airfield as "the two-speed Morris", because it was always driven either flat out (60 m.p.h.), when going to work, or nearly dead slow (20-25 m.p.h.), when out for pleasure on the roads to and from the hills and in the glens. I kept it in the minister's garage in Hillside at a modest rent. The garage was set sideways on a sloping lane between two stoutly built stone walls eight or nine feet high, and it took never less than three cuts back and forth to manoeuvre the car into the garage. One Sunday evening, returning late and tired from an active day on the Grampian hills, and finding the steering rather stiff, I finally backed the car into the garage a bit too fast. There was a loud bang and the back of the car went down about 3 inches. I thought I must have run onto the minister's hay rake and burst both back tyres. But inspection showed that what had happened was that the one-brick-thick dividing wall between the garage and the old horse stalls behind it had decided to lean, all in one piece, on the back of the car. The car was strong and seemed undamaged at this stage. So I asked Sandy when he was free the next weekend if he could hold the wall while I drove the car away. He tried, but, not surprisingly, his strength gave out and he dropped the wall, which was shattered into many pieces. A day or two later, I called on the minister and attempted the delicate task of explaining to him that, although the accident was very unfortunate, I had actually improved his garage by enlarging it. I paid him £1 compensation, and we cleared away the bricks and the minister seemed satisfied!

That car enabled me at weekends to have many a healthful outing on the hills, which were ten miles or more inland from Montrose. I camped in the snow up Glen Lethnot one weekend in December as an experiment, to see how effectively one could insulate oneself from the cold ground. That was quite satisfactorily achieved with layers of newspaper on the groundsheet, but it did mean that I did not thaw the ground at all and suffered an uncomfortable night because the ground was stone-hard and knobbly. Another weekend, in October 1938, I had used it to take my little tent to camp under the pines on the Mar estate at the head of Deeside, and from there I climbed Scotland's second highest mountain, Ben Macdhui (4,296 ft or 1,309 metres above sea level). But although I was sure of the weather, which was calm and mild, with high cloud, and although I exercised the greatest care all the way not to twist an ankle, it must have been a rather risky venture to do such a climb alone. Nevertheless, there is truth in the counter-argument that climbing in groups can lead people to go past their own limitations through fear of seeming to be less hardy than the others. Going alone teaches one to know one's own limitations and keep within them. For this reason, it has been suggested that everyone should climb alone at least once. The distance I walked on that occasion, up along Glen Derry and then up to the heights by Glen Etchachan and Loch Etchachan to where the grass and heather gives way to a vegetation dominated by dwarf birch and dwarf willows, and then back, would have been close to twenty miles. I passed a few isolated stags grazing but saw no other persons until I was back at my camp. I had been kept awake the night before the climb by the barking noise of rutting stags. The long drive home, after packing up the tent and having a high tea in the hotel at Braemar, was unforgettable with the clear moon above, as I drove through the great Deeside pine forests and over the hill by the Cairn o'Mount road into Angus.

Another time, during a summer visit of my friend Dennis Conolly, we drove to the head of Glen Clova and took the tent up above the 3,000 foot contour on the summit ridge of Lochnagar (3,789 feet/1,155 m. above sea level), ate a hearty steak supper and next day walked on a couple of miles along the high ridge to the top. Walking on these hills, one would be sure to put up one or two grouse with their croaked calls of complaint: "back, back;" "go back, go back", and on the higher levels ptarmigan with their plumage turning white in winter. Sometimes in the glens one came across capercaillies (formerly spelt "capercailzies"), even occasionally a whole family of the chicks, in the woods. On one occasion, I was able to admire a great cock capercaillie with his impressive size and handsome, contrasting plumage - mostly a dark blue-grey, but with patches of white, and brown wings, and bright red around his eyes, roosting on the side branch of a pine tree at the edge of the wood, beside the road in the Glen o'Dye.

It was during these years at Montrose that I surprised myself by having - as far as I can recall - my one and only experience of sea-sickness. The herring and salmon fisheries, the latter carried on with fixed nets attached to roughly 8-foot high poles on the beach, at right angles to the shore-line, were an important part of the life of the region, and I was keen to experience a night out with the herring drifters. I had arranged to do this, and the surface of the sea was oily smooth on the night chosen apart from a gently rolling swell only a few feet high. We left Montrose harbour between 6 and 7 p.m. with me seated in a corner of the wheelhouse where I could see everything and would not be in the way of the serious work with the nets. Below me I could see a blackened kettle which had been charged with not only water, but also the tea and a tin of Nestle's sweetened condensed milk were in before we sailed. I was comfortable enough through the evening, watching the boats and the preparatory work with the net. But around midnight the net was hauled up bit by bit, and one of the fishermen kindly produced an enamel mug and filled it with some of the milky, very over-sweet tea. I thanked him and took it, mainly to show appreciation and be polite. I soon began to feel ill and within minutes I was sick too, but did not feel any better afterwards. And so it continued until the voyage was over, despite the smooth shiny surface of the sea.

    The Gray's household was a grand one for me to be in and we had many a harmonious chat over coffee beside their fire, a regular family event after lunch. We were so engaged when the radio was reporting Hitler's march into Austria in 1938, but the conversation was often about their journeyings to farmer friends in Angus and Kincardine and memories of the life of the countryside. We picnicked together beside Loch Lee up Glen Esk, and five or six times a winter I travelled on the coach with Mrs. Gray and her friends to the whist drives and Scottish country dances arranged by the Women's Rural Institutes in various village halls. And when, after only two years, I had to leave on posting to Donibristle airfield beside the Firth of Forth, Mrs. Gray said I must always regard their house at Oakbank, Hillside as my Scottish home. I was by that time very fully "dug in" to the life of the village and had many friends there. The language of the Gray's home was Broad Scots, and gradually I found myself picking it up. There was one member of the household, a Downs-syndrome (mongol) lad, Frank Smart, of about 30 at that time, who worked for old George Gray in the garden and whose language was a particularly uncompromising Angus Doric. He had been permanently "parked out" with the Grays by his well-to-do parents. Each evening he asked "What time are ye yokin' (i.e. starting work) the morn', Mr. Gray?" And there were other characters round about who provided extra examples of the native speech. Two old ladies, sisters, from Aberdeenshire, the Misses Barrie (pronounced "Bahrrie"), who lived together near by kept us amused by some colourful incidents. When one of them called one day, she sat down near a big old photographic portrait fixed on the wall of Mrs. Gray's father, and said as she looked at it , "Your faither's bin like wor (our) faither, Mrs. Gray. He's hai-en side-fuskers (whiskers)". Consternation was caused, briefly, among the Barrie sisters when some well-meaning person gave them a present of two bottles of champagne. "Fit (what) wid oss be dae'en wi' twa bottles of champion? Oss jist poured it doon the sink", was the account we were given of the episode the next day by the older Miss Barrie. And one time, when Frank Smart went with a message for Mrs.Gray to the Misses Barrie, one of them offered him a piece of the jam sandwich cake that she had just baked. So Frank took the cake, and started eating it at once, leaving the piece all alone on the plate! These folk provided us with an inexhaustible fund of stories, but of course we never laughed about them in their presence. Another bit of the common tongue was that the time was always told, by everyone in the region, in a way that is widely current around northern Europe: "half one" means half way to one o'clock i.e. half past twelve (actually pronounced "half een" in Angus and "half ane" in Aberdeenshire). Similarly "half twa" means halfpast one etc. (cf. Swedish "halv två"). I was also a regular member of the Scottish Country Dancing class that met once a week. One of the girls of the village, Jessie, who attended it had been a nanny in Kensington and was completely bilingual. The class also included doctors' wives and an occasional young doctor, as well as other ladies from the staff of the mental hospital on the hill.

Through all these contacts I quickly became very fully integrated into the life of the village, though I did not have contact with the local "upper crust", the "residenters" as Mrs. Gray called them, largely Episcopalians, whose focal point was the Episcopal Church in Montrose, and whose minister lived at the top of the hill in Hillside. They evidently lived on a different income level from the rest of the population and their social pursuits were quite different - resembling more those of the gentry in the English shires.

Among the friends I met through the dancing class were the village dominie (schoolmaster) and his wife (Bruce and Nan Williams). Many a bright evening I spent in their schoolhouse, chatting by the brilliant, but all too glaring, light of the Tilley lamp. Bruce had been in the R.A.F. and his wife also was a lively conversationalist. I took their children sledging when there was snow, and we had an occasional all-day outing with our bicycles right up Glen Esk. Another time, when I had driven them by car to the head of Glen Clova and had gone beyond there, walking on one of the hill ridges above Glen Doll, I was teasing them by contesting the reality of wild white heather, when I found myself actually walking on a patch of it. Nan Williams's old father, George Yorston, from Wick, himself a retired dominie, then living in Montrose, was a memorable character with tales of his own great tramps over these hills at the beginning of the century. He had several times crossed right over the Grampians from Clova and Glen Esk to Deeside and back. On one of those crossings he had had a nasty encounter with a wild cat in a hay barn where he was sleeping. In the nineteen-thirties there was a very good inn at the head of Glen Clova, at Milton of Clova, as it was still known, though I think milling had long ceased. The place was a favourite resort of walkers from the towns who could motor so far, take an afternoon walk, and get a very good tea with baps and scones.

There were also more than a few sad reminders of the Highland clearances of a century and more earlier to be seen - many of them still there

At Home in Scotland

- in the shape of abandoned, stone buildings, farm cottages, crofts, and bothies with only some of the walls or even only foundations still there.

In February-March 1938 I spent a few days walking and hitch-hiking the road by Huntly, Elgin and Inverness, past Dornoch and Golspie to Caithness for a short stay in Wick and to visit Thurso and John o'Groats. It was generally dull, dry, mild weather, though there were glints of sun, and the sea was blue at the north coast, allowing splendid views across Pentland Firth to the red cliffs of Hoy and the islands, as well as about Duncansby Head, the north-east point of the mainland, and those fantastic tall pillars of red stone, the Stacks of Duncansby. In Wick I had an introduction from Bruce Williams, himself the son of a former minister in Wick, to Harald Georgeson who, despite the Norwegian spelling of his name and his own and his son's authentic Viking colouring and appearance, did not speak Norwegian. They told me in the very first conversation "We are not Highlanders here, ye know, we are Norse. The Gaelic was never spoken here. It was always a point of honour with us not to know the Gaelic." The old man was Consul General for Norway and had been made a Knight of the Order of St. Olav by the King of Norway. And it was quite noticeable in the appearance of the people, and in their speech, as well as in the place names (like Berriedale and Helmsdale) of the northernmost part of the mainland that one had crossed an historical frontier thereabouts.

Working for the RAF at Montrose, which lies on the North Sea coast, just east of the main block of the Scottish Highlands, put me in a position to start building a reputation as a weather forecaster by making use of what is probably the most reliable mechanism in atmospheric behaviour, the shelter effect and, in this case, the down-draught on the lee side of hills. Eastern Scotland between the Firth of Tay and Moray Firth enjoys a frequency of clear sky and very long visibility which is largely due to the repeated operation of this effect upon the prevailing westerly winds. It was never more clearly demonstrated than in March 1938, when the winds were from the west on every day of the month. Glen Quoich in the western Highlands had fifty inches (over 1,250 millimetres) of rain in that month, owing to the abundant condensation of the moisture in the air from the Atlantic as it was uplifted over the western slopes of the Scottish hills. And Braemar in Deeside, under 80 miles east of Glen Quoich, but east of the mountain mass of Scotland and where the west winds would be descending from the hills, that whole month produced just one fifth of an inch (5 millimetres) of rainfall. The process of abstracting the moisture from the air in this way also produces a rise of temperature - the effect known as *foehn* in the Alps - accounting for some remarkably high temperatures over the low ground at the foot of the mountains at their eastern and northern sides.

Occurrences of 27° to 28° C (80° to 82° F) with strong or even gale force westerly winds may be experienced in the area around Montrose and Aberdeen, while on the north side, about Elgin, the south-west winds may bring temperatures as high as 15° to 16° C (or 59-60° F), even in mid-winter. This reliable effect was most useful in answering the young officers' inquiries in matters that most greatly concerned them: "I am going away for the weekend: where shall I go?" and "What will the weather be like at .....?" As long as they were staying in Scotland, or other mountainous districts, the foehn effect would provide reliable answers *except* when the winds were too light to produce a distinct sheltered side. But often the young officers only wanted "to go to *town*" (meaning London, in the supercilious usage of the fashionable upper class).

There are other memories of those years, for instance how on 25th January 1938 I had a date to take a girl, daughter of the headmaster of the North Esk School in Montrose to the pictures. But, on the drive in from Hillside village to meet her at 6 p.m., I spotted from the car that an exceptionally grand display of the northern lights (*Aurora borealis*) was developing. So I told Helen at once: "We can't go to the film after all. This is something quite exceptional." So instead I drove the car to the top of a frozen grassy hill with an unlimited view of the sky, above Laurencekirk, a few miles north of Montrose, and watched what still ranks as one of the finest auroral displays I have ever seen.....while poor Helen chain-smoked cigarettes in the cab of my Morris car. It seemed as if the whole sky was flickering with continually changing colours moving up great shafts of light, which converged like the spokes of an umbrella, to meet in the auroral crown near the zenith. This went on for an hour or two and at its strongest phases the crown was actually a little south of the zenith: a fascinating continually changing display in delicate colours, but evidently not conducive to romance. I doubt greatly, however, whether things would ever have come to romance between us.

There had been an intense calf-love episode during the previous year with a sprightly auburn-haired girl whom I met in the house of a Quaker friend in the south, and when it turned out that Joan would very soon be returning to university at St. Andrews, where I was for a brief week or two in lodgings before moving on to Montrose, it became inevitable that she and I would see a good deal of each other. There is no doubt that we were both in the grip of a very strong physical attraction that hardly brooked any delay, although in the end it came to nothing. She was very religious in a high-Church, high-Tory way that was quite common in England, but from which I had already suffered more than enough (because my father was keenly using it as a sanction for his dictatorial behaviour, including support of nationalistic activities). A life together with Joan R...... would certainly have been an endless succession of squabbles and deep disagreements. Parting was a painful break for both of us, but I have always been thankful that I recognized the dangers ahead in time. The break left me scarred for some years, and I believe that unfortunately the same was true for Joan. But by January 1938 I was certainly more fascinated by the great display of the northern lights than by any new girl-friend.

It had been experiences on yet another partly hitch-hiking holiday to Norway in the summer of 1937 that led me to a clear understanding that the attachment to Joan absolutely must cease. Being now in regular employment with the Met. Office, I had had to put in for my holiday leave at a definite date. This created difficulty, because the cargo ship that my old friend, the broker Leiv Olsen, had arranged for me sailed from Newcastle a couple of days early. I encountered this problem several times, because even those most closely involved with tramp ships can seldom be sure when loading will be completed and they will be ready to sail. When the bad news came in a telegram from Bergen, I set off on the intended day, hitch-hiked to Aberdeen (one lift in a car and one in a lorry, taking 1 hour 45 minutes for the 40 mile journey) and walked round the docks. Nearly all the ships in port were bound for India and the Far East, but there was one lovely white and blue painted luxury yacht - a motor driven vessel, the "m/v Radiant" which I was told belonged to Lord Iliffe of the *Daily Telegraph*, who was taking a party of friends next day to Bergen to fish in Norway. So I immediately went back to Montrose (only 50 minutes this time) to fetch some identification papers showing I was a meteorologist with the RAF at Montrose, and returned with them. I slept - not much actual sleep, as I was too excited - that night under a hedge outside Aberdeen at Bridge of Don and then went to the Caledonian Hotel and had a good plain breakfast for one shilling and sixpence. Afterwards I walked down to the quayside, but Lord Iliffe had not yet arrived. He and his party arrived in two Rolls Royces at 11.15 a.m. I was nervous about my approach to Lord Iliffe after a conversation I had had over the ship's rail the previous afternoon with an understeward, to whom I had said I would ask to come too, with my own sleeping bag and the food in my rucksack: "You never know your luck", I had added. The steward replied "I can tell you your luck right now....Why they couldn't afford to have a stranger with them.... they're millionaires." So I stood on the quay beside the yacht and gave the steward my card with "Meteorological Officer, RAF Aerodrome, Montrose" written on it by pen. What suspense as I waited for results! And what amusement among the crew! The skipper himself gave me the wink and in under a minute I was invited on board to speak to Lord Iliffe. The little gaggle of onlookers stood back amazed to let me get to the gangway and go up. I told Lord Iliffe about the bad luck that caused me to miss the passage that had been arranged for me. His Lordship said I had an honest face and it would be all right. I told him I had my own food with me, but he said I need not worry about food while I was with them. I was introduced to his friends as "our stowaway" and was told to work my passage by forecasting sunshine. But a few hours later, still in bright weather but on a rough sea, I was dining alone with Lord Iliffe, as all his friends were seasick!

When we reached the approach to Bergen, I learnt to my embarrassment that my obliging hosts had not really been going to Bergen. The weather was now overcast and drizzly, with the light fading prematurely, when we hailed a Customs Officers' launch in the offing, and it was arranged that they should take me right into the harbour. They were friendly, took me to the quay and showed me the way to go, and simply asked me to take my passport and show it at the police station the next day. I met my friend, the ship's broker, Leiv Olsen, accidentally on the street as he was walking home from a football match.

While in Bergen this time, I paid a visit to the famous weather forecasting centre, Værvarslingi på Vestlandet, at the Geophysical Institute. At first there were only assistant staff there on duty and they said "We will not try to speak English. You just speak Norwegian." Later on, one of their leading, internationally well known forecasters, Dr. Sverre Petersen, came on duty and I had a good first meeting with him. He had just married a new wife, his fourth. This one was English. He was encouraging to a new young meteorologist. In the war he worked in England, but I came to know a colleague of his from north Norway, who also escaped to England, much better. That evening I left Bergen with the s/s Nordfjord, one of the regular passenger ships sailing up the coast. I slept heavily and nearly missed the call at Måløy outside Nordfjord where I was to get off. I did photograph the formidable cliff, Hornelen, which at a sharp angle in the channel between island and mainland rises from the sea to 860 metres (2,821 feet) so close to the water's edge that ships are forbidden to sound their sirens for fear of bringing down rock. But there was no time for any breakfast. I stayed with the ship into Nordfjord, the northernmost of the three great west coast fjords that strike into the interior, and finally went ashore at Sandane at the head of a southern branch, after standing in the bows for a while during the unloading of fish and tin plates and wooden roof-trusses for houses, chatting with a young Englishman from St. Johns College, Cambridge and watching the folk standing by, a beautiful 16-year old blond girl with her bike just standing in the sun looking at us and younger boys and girls undressing and bathing naked in the fjord amid shouts and laughter. We readily agreed how contented and at ease the Norwegians in the countryside are. It may well be that an evident lack of tensions has to do with their free and easy - at least relatively - attitude to sexual problems. Young British girls that one saw in Norway at that date were noticeably shy by comparison. But it has become obvious in later years, as attitudes have loosened up in both countries, and the statistics of broken marriages, suicides, and abandoned wives and children have risen until they became higher in England than in Norway, that there is a limit to how far free and easy attitudes can go without causing new troubles, among which AIDs and AIDs-related diseases now have to be included.

I boarded a country bus from there, planning to make my way south to the great Sognefjord. But we had not gone for more than an hour before we came to a stretch of the road where there were occasional glimpses of the great ice-sheet, Jostedalsbre, the biggest surviving relic of the last ice age in Europe, which still sprawls along the tops of the mountains between Nordfjord and Sognefjord for forty to fifty miles (about 70 to 80 km.), spilling many individual glaciers down over 6,000 feet (2,000 metres or more) at several points. The ice has shrunk back a great deal since its maximum extent in the mid-eighteenth century or rather earlier, and the glacier tongues in the valleys have thinned. Great was the alarm among the farmers and the pleas for tax remissions in the times when the ice was building up and advancing. As there were only six or eight people on our bus, the driver was telling us now and again about points of interest. And at a lane end at a small place called Skei, he told us: "At the head of that lane, there is a farm, Lunde, where the farmer is a guide who takes people across the ice." I thereupon asked if I could get down from the bus and headed off up the lane with my rucksack.

So, in a gruesomely unprepared state, I set off to cross the ice-sheet, which was only just over a mile (2 km.) wide at this point. On my feet I had only my smooth-soled London walking shoes, but I thought I would be all right on the rope with the guide; and so I was, but he did have to pull me up out of every crevasse. I had to spend the night at the farm. We started off at 5.40 a.m. next day. The ice was in terrible condition in that long, warm summer, melting fast, running with water and at its slipperiest. On the climb up over steep grassy slopes to the ice Hr. Andreas Lunde complemented me on my speed, but once we were on the ice it was a very different story. And I was ridiculously laden, having bought a home-weave bedside mat from the farmer's floor. This was rolled up, hanging beneath my rucksack. But we had got across by 10.10 a.m. The guide told me his father remembered bears hereabout and killed one down in the valley. I paid him and then set off down the 5,000 feet of equally steep slopes on the southern side. The effort of trying to keep my balance on the ice, and avoid slipping, had been so great that my legs kept on folding up under me on the way down. I got a lift on the valley road halfway to Fjaerland, but had to walk the last 3 km. when the car turned off on a side road. The car was carrying ladies who had been on the bus yesterday. I was dog-tired when I reached the hotel at Fjaerland. And there I got a hot bath, even though it was about 11 a.m., and began to feel less stiff. A pretty girl had taken away my shoes and stockings to dry and returned them later without charge. At that time there was no road to Fjaerland, and it was a strangely lonely place, isolated except for the fjord ships and dominated by the ice-capped mountains blocking the head of the valley.

From there I returned to Bergen by boat, visited the Geophysical Institute and met Dr. Pettersen again and his assistant on the weather forecast duty, Alf Maurstad, who was learning Spanish and going to take up a post in the Argentine Meteorological Service. After supper in the Holberg rooms, I spent the night in the youth hostel. Then, by 9 a.m. next day, to the station and the Oslo train, which was to leave at 9.25 a.m. I was to learn something about myself on that train. Before the train left a rather tall, very handsome blond schoolgirl or student in smart clothing sat down opposite me at the same table. At first, she wanted the window down, and I had to help her arrange it. Soon, she wanted the window up again and, not long afterward, down once more. It was a brilliantly sunny day and one soon felt warm behind the glass, and other passengers started pulling the blinds also. Once most of the blinds had been adjusted, I settled down to work out the crossword puzzle in the *Hjemmet* magazine that came in the daily newspaper, *Aftenposten*. I believed I was very much in love with Joan, and had avoided a fair number of conversational openings provided by the smart schoolgirl. But now she went down the train and came back with her own copy of the magazine with the same crossword puzzle and started to do it, just on the other side of the same table opposite me. I was not proof against that, and soon we got talking. The whole of the rest of that day I was almost inevitably talking, talking Norwegian, with the girl and with an old seaman from the Hamburg-Amerika Line who came and sat beside us. His crew mates, he told us, were nearly all Norwegian. By the time we arrived in Oslo Joan's spell over me had been completely broken, and I was met at the station by Tunny Muller looking prettier than I had ever seen her and accompanied by her sister, Dorothea. I was deeply shocked, and indeed worried, at the discovery of my fickleness but also, more gradually, became fully aware of just how fundamentally ill-suited to each other by background and outlook Joan and I were. Ten long years were to go by, during which I began to doubt whether I would ever come to love anybody so deeply and firmly that I could be true to them. During that time circumstances uprooted me from Scotland, and it turned out that I came to know life in southern Ireland. The war separated me from Norway, where I had clearly struck long-lasting roots, so that I am often inclined to feel more at home there than anywhere else. No doubt the facility I found I had with the Norwegian language was a great relief and encouragement to me after my dismal failure with French. As it turned out, this was to be a real help to my career both during the war and after.

CHAPTER 9. A VISIT TO ICELAND

For my main holiday in 1938 I planned to sail to Iceland - it was before the days of a regular air service - and, by changing boats, to go all round the coast of that country before returning to Leith. The plan appealed to me both because of my respect for what the Icelandic population had achieved down the ages, with their oldest parliament in the western world and their very high concern with education and literacy, as well as because it seemed to me important to sample the weather processes affecting everywhere adjacent to the projected trans-Atlantic air route. Iceland was particularly important in connection with weather forecasting in Scotland because weather systems coming from there could reach us in under twelve hours. I sailed with the Iceland Steamship company's m/v *Lagarfoss* in the late evening of 22nd September 1938 and was politely seen off by Mr. Cairns of Cairns & Co., Iceland Steamship Co. There had been rumours of war that day and Mr. Cairns wished me a very happy trip, adding that "the waar'll likely be over afore ye get back". Very soon after we had sailed I had to deal with an emergency in my cabin, which was very high up above the top deck. I had noticed that someone had left the porthole wide open, but in the prevailing calm weather that did not seem important. However, as soon as the ship emerged from the Firth of Forth into the North Sea we encountered some swell, and evidently within minutes a wave came right in through the porthole and spent the next half hour chasing itself round and round the cabin. It took longer than that before I had got all my things dry!

Four days later, while northbound off the east coast of Iceland, I received a telegram sent to the ship saying "Return Montrose immediately. Aeronautics Montrose." The captain called me to his cabin and, looking worried, said: "As far as I can see there is no way you could get back to Montrose quicker than by proceeding with your planned trip." That was to continue with the *Lagarfoss* to Akureyri, change ships there and change again later in Reykjavik, calling in at the Westman Islands on the way. There an inhabitant pointed out to me the all too obvious shape of the hill above the town, a volcanic cone if ever there was one. He said "That is an extinct volcano, you know." I replied by saying "I would not trust it". And, indeed, thirty five years later, in January 1973, this "extinct" volcano, Helgafell on Heimaey, erupted again destroying a large part of the nearby town and harbour. Sufficient warning was given and there were no human casualties.

The trip around Iceland was a fascinating and instructive one. Our first call was at Djupivogur, a tiny farming community on rocky flats at the mouth of a little fjord in the south-east of the country. The houses were built of wood as in Scandinavia proper, but there is no wood for building grown here. The roofs were bright red, as elsewhere in Iceland, or made of turf. We passed an island in the inlet called Papa-ey (pope's island), where the first Irish monks came, probably in the eighth century, and settled, it is said, some seventy-five years before the first Scandinavians reached Iceland.

The ships on which I was sailing round Iceland were performing the essential public service for passengers and freight. They went anti-clockwise about the country, an island a fifth bigger than Ireland. We called at twenty-seven places, nearly all of them tiny, remote settlements. Most of the calls were only for an hour or two, and for most places there was only one service a month. It took four days from Leith to Djupivogur and four more working round the coast to Akureyri, the main town in the north, halfway along the north coast. After three days there, we headed on farther west and put in later the same day at Siglufjörður, a whaling station deeply tucked away at the head of a mountain-girt fjord which seemed to be entirely filled with the unpleasant vapours and stench of the whaling. We never got rid of that smell again until we reached Reykjavik a couple of days later, because all the passengers who came on at Siglufjörður had their clothes thoroughly impregnated with the whaling vapours. It was a very dull, calm overcast day when we were in Siglufjörður, and although I climbed a path far up the hillside while our ship was loading at the quay, there was no escape from the odours that way: it seemed that the atmosphere was imposing a lid on the whole fjord valley so that it was like being in a box, open only at the north end and with no air movement in or out.

But if that was the ugliest place we struck around the Iceland coast, others were far pleasanter. There was a thin covering of new snow- ever since 1st October - along the upper third of the hills here in the north, but the open air swimming pool on a little hill beside the fjord at Akureyri was still well patronised by young and youngish people. Like many swimming pools in Iceland, it was filled with natural hot water, of volcanic origin, and the temperature in the bath was kept at about 28° C (82° F). Icelanders whom I talked with asked me whether it was really safe to bathe in the sea in Scotland, where the inshore water temperature in my last swim at Montrose before coming away had been just 13° C (55°F). One very notable characteristic of the Iceland scene, and one that the Icelanders are rightly very proud and fond of themselves, is the blue of the distant views. In the characteristically very clean, dust free Arctic air, hills and headlands within the line of vision are seen at distances up to well over a hundred English miles and, even at shorter distances, acquire an opalescent blue colouring like a rather darker version of the blue sky itself. This is attributed to the scattering of the sun's light by the molecules of the very pure air. It is best seen when not looking directly towards or away from the sun. Views of a coast from some miles off shore are commonly distorted by mirage effects, here known as *fata morgana* which may undergo gradual changes of form while one looks at them. This is due to reflections and bending of the light rays by a glassy calm cold sea surface under warmer air. As generations of sailing directions and pilot manuals dealing with coasts in high latitudes have warned, even familiar views may be made unrecognizable. Raised images of scenes along the horizon, and inverted images, are often involved. Combinations of right images, raised images, and inverted images occasionally produce weird shapes. Examples seen along the north coast of Iceland included a valley converted into a huge tunnel with the actual pass repeated above: and all this in beautiful sunshine. The phenomenon is not quite confined to high latitudes, and I was subsequently to see similar raised images of the shores of the wide river Shannon in south-west Ireland.

Objects which should be below the horizon may be brought into view in their raised images: on 29th September 1938, while cruising westward along the north coast of Iceland near the entrance to Eyjafjörður, we were able to see the outlying northern island, Grimsey, 35 miles away although its highest point only reached 110 metres above the water: it first appeared upside down, then right side up.

A chat in my cabin as we were approaching Akureyri gave me my most striking experience of the Icelandic reverence for education. I was talking with a youth from Borgafirði on the east coast who was in his fifth year at the school in Akureyri where he would be until the end of May 1939. Already he spoke almost perfect English, without accent but rather slowly. He knew much more Latin than I could remember, and could speak German and Danish in addition to his native Icelandic, though these were not as good as his English. In this next school year he would be starting French. His aim was to travel for a year or two and then be a teacher. If he succeeded in getting a scholarship from the school he would hope to go to a foreign university. Otherwise, his plan was to travel as a seaman or a steward and save until he could go to some foreign university. His parents, being small farmers, were too poor to pay for his schooling. They would not manage to employ a worker on their farm to do the work he used to do. For more than a year past he had kept himself and paid for his schooling. The most hopeful line of support was to get employed by the state on road-making in Iceland, but the state itself could not afford to take on all who would like to earn the money.

At Akureyri I saw more trees than anywhere else in Iceland. They were all in a plantation of pines that made a park a mile or two outside the

A Visit to Iceland

town, in the lower part of a valley coming down from the interior. The sand banks in the river were made of black volcanic sand. Paths had been laid out under the trees, but one would need to crawl under the lowest branches to get along them. Nevertheless, for the Icelanders this wood was an achievement. In other parts of the island, one saw only single trees, often rowans, planted in gardens with neat white picket fences; but growth of the trees seemed to be always limited to the height of the roof of the house. In the Akureyri district, and possibly elsewhere in north and north-east Iceland, the climate has some warmer moments. Higher temperatures have been recorded than elsewhere, up to about +25 °C occasionally. Despite the area's northern position, there is shelter from the cold winds on the great stormy ocean to the south. But probably most important is the foehn effect, the condensation heat passed into the air when its moisture is abstracted as it is lifted to cross Iceland's high interior: this warms the south and south-west winds before they reach the north-facing coastal region. The same process warms the north-eastern Scottish climate about Elgin and the north-facing coast lands of Moray Firth.

Our calls in the north-western fjords were made in very cold, calm, overcast weather, which emphasized the remoteness of the little places, Isafjörður, and others. These mountain-girt fjords open only to the north-west, to the Denmark Strait between Iceland and Greenland.

We reached Reykjavik, Iceland's capital, in the south-west of the country on 6th October. At that time, it was a town of about 33,000 inhabitants in a country with a widely scattered population totalling just 120,000. Old ways were still persisting. I was told at that time that the old custom whereby a girl must not put up her hair before she was either married or thirty years of age was still generally held to, whereas once she married or passed the age of thirty she must make up her hair and wear a black lace cap. Certainly many of those caps were still seen on young women of about that age. The Icelandic way with family names is still very much adhered to and, indeed, reinforced by law. Surnames are not used in the international way and really do not exist. If Harald has a son Jon, he is known as Jon Haraldsson. His daughter Helga would be known as Helga Haraldsdottir. And this Jon's son Harald would be Harald Jonsson. This custom presumably comes right down from the beginnings of Icelandic society and may reflect an experimental attitude to marriage in the earliest days when exposure to the risks of life in such isolated communities in a wild and lonely environment was particularly dire.

My first call in Reykjavik was at the Veðurstofa (Weather Room or meteorological institute), where I had the good luck to meet the chief forecaster, Jon Eythorsson, who had been trained and worked as a forecaster in the Norwegian Geophysical Institute in Bergen for six years from 1921 to 1926. This was the first visit he had received from a British meteorologist in his twelve years experience in the Iceland service. He was particularly interested in my investigation of the Scottish North Sea coast fogs (haars) and their association with the wind circulation within the coastal sea breezes. He himself stressed the importance of the temperature inversion (higher temperatures above the fog). Around Iceland the fogs over the cold sea were sometimes so shallow that ships' masts stood out above them. He had a peculiar difficulty in training Icelandic observers to report fog: they were so used to almost limitless visibility that many would report fog if the visibility was less than 10 km. (6 miles) or if there was cloud on the hills. He emphasised the great importance to Iceland's weather of variations of the ocean surface currents. Iceland experiences water of Gulf Stream origin, which is warm for the latitude and brings the jellyfish with it and which also affects the drift of the herring. The current ordinarily approaches western Iceland on a broad front: it then branches towards the north-west and west and sends a smaller branch east along the north coast of the country, but this encounters a branch of the East Greenland (cold) current coming from the north and north-west. The outcome varies from season to season and from year to year, being affected by the varying volume flow of the Greenland current. From this results a southward flow along the east coast of Iceland and westward along the south coast, which may consist more or less of the cold or warm water from time to time. Recent studies suggest that the increased volume of cold water off eastern Iceland in the 1960s and 1980s had to do with increased supply from the Canadian Arctic rivers after increased downput of rain and snow over the Mackenzie River system. An earlier case in the eighteen-eighties even brought tongues of the polar sea ice itself back to the Faeroe Islands in 1888, and there was a rather similar reappearance of the ice on those waters in 1968 and 1969.

Jon Eythorrson was so good as to spend most of that day with me and invited me to watch him performing his forecasting duties, including accompanying him with his charts to the studio where he spoke the forecast over the Iceland radio, without written notes, to the public and to the fishing and sea-going interests. His detailed weather charts, with the barometric pressure lines at a close interval, showed many local effects on the wind-flow caused by the ruggedness of the Iceland terrain, with many hill ridges repeatedly forming barriers to the wind flow which would be forced to go round them, producing localized pressure minima (troughs and cyclonic centres) in the lee of the barriers. This is a feature of Iceland meteorology which is the more recurrent because the air is cold and not so easily forced over the obstacles. I was also invited to share the high-quality catering products - coffee and elegant cakes and biscuits, served to the weather bureau staff on duty.

There are exceptions to the normally very clear air in Iceland, when a dust haze is picked up from the lava deserts of the interior and when smoke (e.g. from fishing vessels) is trapped in the lowest air under a strong inversion of temperature (which means that in such cases the smoke-laden air has not been heated enough to rise through the warmer air above).

From the weather service building I went on to look at the new, and at that date uncompleted Catholic cathedral, which was also on the high ground, a striking building in ferro-concrete. I attended Vespers there so as to be able to add my prayers to those of the good people of Reykjavik, who were continually voicing in every conversation their anxieties about the threatened war.

I was also driven, at the kind invitation of one of the seamen from the ships I had sailed around the coast on, to Thingvellir, some thirty to forty kilometres (twenty miles or more inland), where in a fairly broad valley setting the early parliament, the Althing, met from AD 960 onwards. It is a landscape created by volcanic activity, with vertical sided clefts and trenches, a mountain massif in the background, and a big farm beside a largish lake in the valley before it. The members of the parliament used to gather in one of the broadest clefts and were addressed by speakers from an elevated position alongside.

I managed just one more short holiday in Norway, my last visit there, and to my friends in Oslo, before the war closed in on us all. In this case the object was first steps in skiing. In March 1939 I sailed with the m/v Blenheim from Newcastle to Oslo, where again I stayed a couple of nights with my friends, the Müller family, and their daughter, Tunny, advised me on my choices of skis and ski clothing. They then saw me off on the train to Tretten station near Lillehammer, a journey of several hours into the broadest part of the country, heading north into the great long Gudbrandsdal valley. It was an unforgettable journey into the world of deep snow and frost, with tried and tested tracks for wheeled vehicles across frozen lakes and the frozen river, marked out by torn-off pine branches. In every town along the way there were people taking their shopping home with the aid of a *sparkstøtting*, a sort of light wooden chair on runners (literally having its own short skis). Everybody was used to the conditions and life went on. From Tretten a taxi with steamed up windows drove me and others up the winding road through the forest to Gausdal High Mountain Hotel. Here and there along the way, we passed a horse and driver pulling felled pine trunks held together by a chain wound loosely round them through the snow to the roadside, a task since mostly mechanized. At Gausdal I had my introduction to skiing with a few lessons on the beginners' open slope and one or two very cautious outings along the simplest, marked trails. During the first five days of my stay mist covered all the hill ground, while longer and longer ice needles continually grew longer on the trees. By the sixth day the wind had changed right round from SE to NW, and although the wind was still light the mists had all cleared and revealed a marvellous countryside sparkling with white ice crystals everywhere under the very clear blue sky. Weird rounded shapes made the bent-down snow-covered trees look like trolls and hunch-backed witches with the ice crystals, many inches long on their other sides, irresistibly suggesting the noses and fingers of the hunched figures. It was easy to understand how the tales of such creatures in the Norwegian mountains had arisen. That night, at the end of the evening meal in the mountain hotel, the young host stood up and made an enthusiastic speech which received general acclaim and obviously expressed everyone's feelings. "Yes, yes", he began "*Gud skje takk* (God be thanked). The fairy tale land is here again...." It was a memorable outburst of natural religion, before the war clouds closed in and took their toll.

CHAPTER 10. THE NATIONS PREPARE FOR WAR

The full social life I had been enjoying in Hillside village was brought to an untimely end by the gathering clouds of war.

Christmas and New Year in the 1938-39 winter were a time of very wintry, snowy weather. Down in England it was a snowy Christmas celebrated with sledging with my friends, the Leaning family, on Hampstead Heath. And back in and about Montrose I met all my friends there skating and sliding on the frozen dams (ponds) and the drainage ditches near the golf course in Montrose: there cars were parked with their headlights on to enable the skating to go on through the evenings. It was during this time that old George Gray told me how the sheep in the Grampians know that the overnight frosts are usually severest in the valleys and on the low ground, so that left to themselves the sheep tend to move up the hillsides around dusk.

At New Year 1939, while visiting my parents in London, with some very welcome financial aid from my father, I bought my first nearly new car. It was a little, open-topped Flying Standard eight horse-power that had been used as a demonstration car by a West End agent for Standards. We paid £115 for it. I took delivery on the 1st or 2nd January 1939 and set off at once northbound for Montrose, a distance of about 500 miles, in pouring rain. Somewhere in the Chilterns the rain began to turn to snow, and the journey from there on was over lying, mostly frozen snow. The little car, like most cars in those days, had no heater. I had to stop occasionally and walk up and down for a few minutes to bring back the feeling in my feet. I night-stopped in a little old half-timbered cafe-hotel in the centre of Penrith - it would have been called a cafe-pension on the continent. Next day, the drive continued in the frost over roads covered with beaten down snow and glazed ice. The most treacherous surfaces were over Beattock summit, where the main road north passes over the Southern Uplands to Abington and the road divides for Glasgow, Stirling and the central Lowlands of Scotland, or Edinburgh. Over Beattock the lorry traffic, as always, was quite heavy. Heading back to Montrose I took the middle choice that led me through Stirling and Perth to the broad Strathmore gap between the foot of the Grampians and the coastal hills behind Dundee. A fine route, all the way through exhilarating scenery, but there had to be quite a few more stops to walk and stamp the feet to prime my circulation. But, within a week or two, I had to leave my well-loved home in Hillside near Montrose and move to an office on the Royal Naval Airfield at Donibristle, in Fife near the Forth Bridge. H.W.L. Absalom of Meteorological Office headquarters in London, who knew how happily "dug into" the local community I had become, but who also knew my work on the haars, was involved in the decision, and was himself the bearer of the news. He was very apologetic but full of good wishes.

I found good lodgings in the middle of Dunfermline, near the Abbey and beside the park, known as The Green, where the barking of the peacocks used to wake us in the mornings. Although I only lived there for under six months, I was unusually quickly integrated again into a lively social life through, almost accidentally, being recruited by the Dunfermline Dramatic Society. It was my first (and almost only) experience of acting, but seemed to go very well and the performance was very favourably reviewed in the Dunfermline Press. I was cast as the stateless Jewish refugee among the surviving passengers on a crippled ship after a fire at sea in J.B. Priestley's play "People at Sea". Through the members of the cast I soon made other friends too. One was a great walker in the Highlands, known to have covered 45 miles in the hills in a single day. Others were members of the Catholic church, some of them Irish. But I found particularly interesting a young Catholic German refugee who had just arrived from Czechoslovakia, whom I drove around the country, introducing him to Scotland. He was a Sudeten German, that is a man of Czechoslovak nationality but of purely German blood, from the Sudetenland, the old German areas of Bohemia just inside the Czech frontier, which Hitler was loudly claiming. This young man used to wring his hands in impotent disgust and distress at Hitler's acts and propaganda. He told me "Hitler will bring Germany to ruin. He claims he is building up and preserving every well-loved German tradition, but his actions will end by destroying it all."

For me myself and my career a most serious personal crisis was looming. It broke without warning about the end of June 1939 when one morning a teleprinter message came to the office instructing me to go for some days temporary duty to an RAF airfield down in England for a practice exercise in (poison) gas spraying. This was something which I clearly was not willing that my meteorology should be used for. So I replied at once and quickly received another message on the teleprinter calling for my resignation, subject to the statutory one month's notice. I immediately signalled my agreement. The resignation demand had in fact come from the deputy director in charge of personnel, R. Corless, in the temporary absence abroad of Sir Nelson K. Johnson, the director. When Johnson returned early, two weeks later, he sent another teleprinter message, asking me to reconsider my resignation and whether, in this connexion, I would be willing to be seconded to the (then newly formed) Irish Meteorological Service to resume work on the forthcoming Transatlantic civil passenger air route, with possible eventual transfer to the service of the Irish Republic. It was a very civilized solution, to which I would have felt obliged to agree quite apart from my pleasure in it. The conscientious objectors in the first World War had, in general, met with no such understanding. Many, I know, had faced much hostility, and I supposed there must still be many people about who would not approve of the solution I had been offered - although I never myself experienced any obvious expression of such disapproval, a fact which made me proud of the British community.

I used my little car and the time available to me for brief further visits to the Grays at Hillside, to the Patons in Edinburgh, and of course to my parents. Jimmy Paton's friendship was particularly valuable to me because I was always worried lest, in the long run, my mathematics might prove inadequate for the meteorological career. The warmth of the welcome that my report of my investigation of the sudden onsets of the North Sea fogs on the Scottish coast, known as haar, which contained a mathematical section dealing with the air circulations, turbulence and buoyancy questions involved, had surprised, as well as gratified me. But such confidence as that engendered was later undermined when a second edition of my paper appeared with the mathematical section re-written by a colleague whom I did not know, and had never met, and who never consulted me over the refinements and revisions which his text introduced.

This experience stirred unease in me and then made me interested in the proposals of the director of the new Irish Meteorological Service - that were freshly advertised a little later - that their scientific staff, to be trained for forecasting duties, should receive a two-year course of training as opposed to the British Meteorological Office's six-month standard course. One year of the 2-year Irish course was to be spent at the famous Massachusetts Institute of Technology in Cambridge, Mass. (near Boston). I was attracted and felt sure I would be much more of a meteorologist, if some day I were to go through the course that the Irish Meteorological Service proposed.

1. Hubert Lamb c. 1975, near the Climatic Research Unit, University of East Anglia.

2. Professor Horace Lamb, c. 1919-20, in Manchester, aged about 60.

3. Hubert and Moira Lamb, at home in Holt, Norfolk, in 1994.

4. Catherine, Kirsten and Norman Lamb, the next generation, in 1994.

5. The grandchildren: Adam, Kate, Archie and Tom, and Anna with Catherine and Kirsten, in 1995 and

6. Ned in 1995-96.

7. The author's mother, Lilian Brierley in 1904, as a young woman of 27.

8. The author's father in 1914, aged 36, Prof. Ernest Horace Lamb.

9. The author's father and mother in relaxed mood, c. 1938.

10. Sunlit birch tree on a mountainside at Gausdal in Norway, showing 5 to 6 days growth of ice crystals produced by a steady SE breeze in hill fog: as seen on 8 March 1939, after a switch of the wind to W to NW.

11. Fjaerland: end of the road and of the valley and fjord, with the Norwegian ice cap (Jostedalsbre) and glacier tongues in view. (Today the road no longer ends at Fjaerland but continues through a tunnel under the mountains and ice cap.)

12. Chamonix valley in the French Alps, from a painting by Jean Antoine Linck in the 1820s when there were long, extended glacier tongues reaching the valley floor, which have since disappeared thanks to the prolonged glacier retreat.

13a. "The fairy-tale land is here again": pine trees encrusted in ice and snow including…

13b. …some in the form of old troll women – Gausdal area, March 1939.

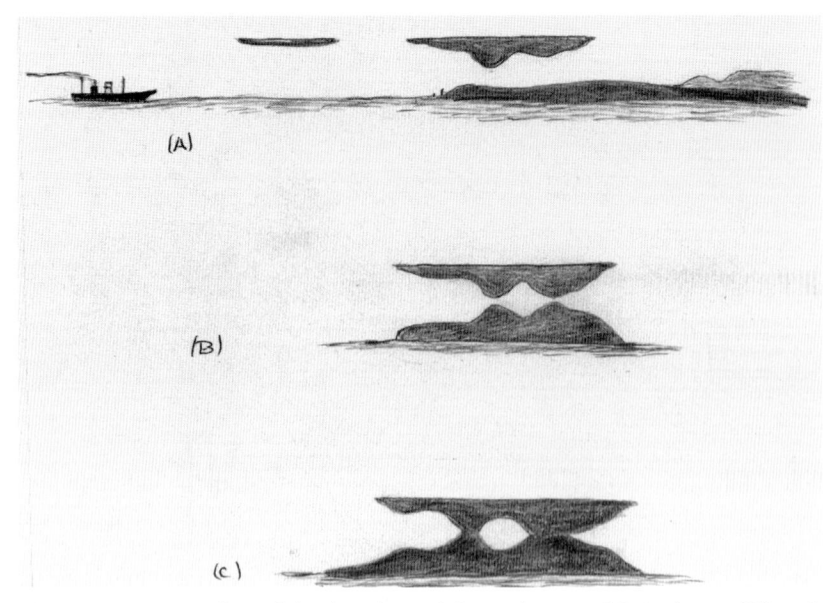

14. *Fata morgana* – mirage deforming views of the north coast of Iceland, near Rifstangi, September 1938.

15. Sailing along the north coast of Iceland, westbound, near Eyafjörður, September-October 1938.

16. Whale-factory ship (Wh/F) *Balaena*, seen in Cape Town harbour, with Table Mountain, November 1946.

17. Strange moonrise in the Antarctic, sketched about 12h. GMT, 15 March 1947, near 64°S 106°E. Wind SW to W, force 5 Bft, temperature -6.8°C, weather clear. (About half an hour passed with the pink drapery continually undergoing some changes of form before it became obvious that the object being watched was the moon.

18. Calm, luminous sea at the edge of the Antarctic pack-ice, under a bright night sky, near 60°S 67°E, temperature -3°C.

19. Smooth swell rumpling the open water surface at the edge of the pack-ice, with wind off the ice and one tabular iceberg: 64 to 65°S near 108°E, wind southerly force 5.

20. "Walrus" amphibian aircraft for whale spotting, catapulted off the factory ship Wh/F *Balaena* over the Southern Ocean, February, 1947.

21. Flensing: deck scene on Wh/F *Balaena* with work beginning on stripping the blubber from the season's biggest (blue) whale carcass, 93 feet (28.4 metres) long by 13 feet (4.0 metres) thick, on 12th January, 1947.

22. Tanker m/s *Norvinn,* moored to Wh/F *Balaena* for the transfer, by pumping, of our oil production. Note the inflated whale carcass used as fenders to keep the ships apart.

23. Tabular iceberg in sunshine. Note the caverns worn in the ice at the waterline.

24. Ice wilderness: profusion of 'bergs and innumerable smaller pieces of ice, after a storm, in 64 to 65°S, near 101°E.

25. Mountainous sea: near 50°S, 25 to 30°E, in the zone of the Brave West Winds, April 1947.

26. Rough water in the breadth of the Southern Ocean zone of Brave West Winds, near 50°S, 25°E, April 1947.

27. The Author with attenders of the Climate and History Conference, in the University of East Anglia, July 1979.

CHAPTER 11. WORK IN IRELAND AND THE NEW TRANSATLANTIC AIR ROUTE

I had never been to Ireland and had some misgivings about how I would take to living in such a predominantly Catholic country. Ninety-nine per cent of the population of Co. Limerick, where the flying boat base was to be were Catholics. My one Irish-born relation, Granny Foot, was of the minority persuasion, a Protestant, i.e. by origin a member of the Ascendancy. Anyway, she had died ten years before. I had heard a little of her family's life in their great house and garden, Holly Park, on the slopes of the Wicklow Hills, near Rathfarnham, Co. Dublin. Quite the "wrong" background really for acceptance by the Irish population. The first of my grandmother's ancestors to arrive in Ireland had taken part in the Battle of the Boyne in 1690 in King William's forces. An uncle of hers had been murdered on his estate in the Wicklow area back in the eighteen-thirties, and all the Foots had ultimately emigrated, mostly to Canada. In the nineteenth century the family had been prosperous tobacco merchants in Dublin, who made their fortune in snuff and, according to reports, carried on with snuff too long, after the fashion had changed. The only living link in my day was through my uncle, Henry Lamb, R.A., the artist, who had married Lady Pansy Pakenham, a daughter of the Labour peer, Lord Pakenham, whose brother, the then Lord Longford, was well known for his travelling theatre players who took live theatre to many provincial towns in Ireland. But I had never met them and had very seldom seen my Uncle Henry or Aunt Pansy. I did, however, start with a general tendency to sympathise with the people who had been unfairly treated by their overlords.

So, one morning in July 1939, I duly went ashore from the Fishguard ferry in Waterford to discover what lay before me. My little Standard "8" car was loaded up to the hood with my bags and coats and shoes. Beside me, occupying the passenger seat, was a wooden chest, that had been my "tuck box" at Oundle and was now full of photographic equipment. A mile out of the town driving up a hill, I was stopped by a priest waving his umbrella before me, who was quite angry when I showed him that I had absolutely no room to offer him a lift. Farther on, going along a street in one of the towns I passed through, with terraced housing on both sides, I was shocked by a white hen dashing out of an open door, dodging between the parked donkey carts and then shooting across the street in front of me. With some difficulty I managed not to hit it. Farther along the way, I came across stray animals grazing on the verge and, once in a while, a beast asleep lying partly across the roadway.

To Ireland and the New Trans-Atlantic Air Route

A couple of years later, I drove an elderly Jewish refugee scientist, Professor Conrad Pollak of the University of Prague, who had been found a teaching post in the Irish Meteorological Service at Dr. Valera's behest, and his wife, to the far south-west of Ireland, and they exclaimed that the experience reminded them of driving down into the Balkans. In those days one became used to horse and donkey traffic passing one on either side of the road. Particularly on Sundays, when the road was crowded with a rush of traps coming away from Mass, they would regularly come towards one on both sides of the road. Luckily, I never ran into any beast, though I did see many animals hurt by other drivers, especially when, after rounding a corner, one came upon an animal, sprawled on the tarmac, asleep, with its length across the road. One dark, rainy, November night, in the first winter of the war, when driving from Limerick to Dublin, I had a nasty shock when with the shiny, smooth road surface glistening in the (reduced) headlights of my car, the road seemingly empty, I became aware of a tinker's unlit caravan at some distance directly ahead of the car. I applied my brakes and found myself skidding straight ahead. That was lucky, because I soon noticed greyish-looking woolly shapes on both sides of the little car and realized that I was skidding straight on through the midst of a small pack of goats that were following the caravan. My luck held out; I did not touch any of them.

The actual first Trans-Atlantic passenger flight had taken place in July 1939, shortly before I arrived at Foynes. On my first Saturday afternoon there, I happened to be off duty and, as the weather was fine, I borrowed the hotel boat and spent the afternoon rowing about the harbour and part of the way reconnoitring round Foynes Island, a half-mile diameter island with a low hill and some high trees in which cranes nested. I knew there was a flight expected to arrive and I kept clear of the taxiing channel between the open river where the aircraft alighted, and took off, and the moorings in the harbour. But I misjudged the wash from the big machine and, after it had taxied past, my boat was stuck, almost high and dry, on the tidal mud many yards nearer the island shore. I managed to climb out of the boat safely into the shiny, waist-deep mud and got enough grip for my feet on a stonier bottom to push the boat along until it floated again. I stayed most of my first year in Foynes at the small private hotel, Ardenoir, where the garden was just a clear patch on the steep, pine-clad hill about 100 feet above the narrow tidal channel that divided the island, with its farm-land, from the mainland. The hotel was run by a Miss Jean Little and her old Scots mother. She soon married the chief British Airways shore representative, Alan Dewdeney, and the captains and other officers of the aircraft used to stay in the hotel between flights. So it was a good place to meet them and learn more about

To Ireland and the New Trans-Atlantic Air Route

the operation of the flights. The American flight crews from Pan-American Airways and American Airlines mostly stayed at a more luxurious hotel also in lovely surroundings, at Adare, some nine or ten miles away to the east, and we came to know them quite well too. The other British meteorological staff had started off at a pub, "The Shannon Hotel" on the main street of Foynes village, but had moved several miles away to the next village to the west. They included Sidney Peters, the very same Mr. Peters who had trained me and the other Foynes colleagues three years earlier at Croydon. We all knew each other and it was a friendly reunion. Among the others was D.A. Davies, who had since crossed the Atlantic many times on a cargo ship, measuring the upper winds by surveying the flight of pilot balloons, and was later to become Secretary General of the World Meteorological Organisation in Geneva. Another old Croydon friend was Stan Proud, who was early withdrawn from Foynes by the Met. Office for service on an Atlantic weather ship and sadly soon to be sunk, one of the first meteorological casualties of the war. There were also at Foynes five or six young graduate Irish meteorologists, the first batch of recruits, forecasting under individual supervision and still hoping to complete their training at the Massachusetts Institute of Technology. Meanwhile, their climatological training was to be proceeded with in Foynes under Professor Pollak. There were also a number of Irish meteorological assistants to be trained as observers.

The forecasting job for the pioneer Trans-Atlantic flights was an extremely responsible one, very thoroughly planned by old Sidney Peters. Each flight forecast involved ten to twelve hours preparation, analysing the weather maps with greatest care and making forecast maps for different times during the flights. Most important of all was the pre-flight forecast discussion, the "briefing", with the captain and officers in the forecasting room about one hour before take-off. It was an exercise in honesty and complete openness about how much and how little of the expected development of the weather situation over the ocean and the destination (and alternatives - "alternates" in the American terminology) one could really be sure of. Also for discussion were in what ways the forecast might go wrong, and how the symptoms might be recognised in flight, as well as the weather prospects at Foynes (in case a return became necessary) and at the possible alternative destinations ("alternates") in case of need.

After the aircraft had left on its voyage, our radio operators kept in contact and our Air Traffic Controller at Foynes exercised traffic control until the point had been reached where responsibility for control was to be handed over to the American side (or to Lisbon or Bermuda in the case of flights on the winter routes).

To Ireland and the New Trans-Atlantic Air Route

Our friends, the original Trans-Atlantic aircraft skippers and crew, and we ourselves, felt it an alarming loss of direct involvement when, in the course of the next following years, with increasing frequency of flights, some mass production of forecast documents, the same for several flights, inevitably crept in. What was surely less inevitable, and was more seriously opposed by the pioneer flight captains and crews, was the change-over about the end of the war from float planes to wheeled aircraft for use only on land airfields. Prominent among the opponents was Captain Donald Bennett of British Overseas Airways, who after the war stood as a Liberal candidate for election to Parliament. The strongest argument was that in most regions there are water surfaces to be found which could serve as ready-made alternative landing sites in case of emergency.

Soon after the war had broken out, in September 1939, all the British meteorological forecasting staff at Foynes were recalled to work in the United Kingdom apart from myself and one other, Paul Brown, who from then on shared the main forecasting responsibility with me. With German U-boats roaming the ocean and military aircraft of both sides overhead, the only weather observation reports from the ocean available to us in Ireland, a neutral country, were the reports of their in-flight observations brought in to us by the aircraft after arrival. "Debriefing" of the crews after each completed flight took on great importance. Even so, the difficulty of keeping going a reliable analysis of the weather systems over the ocean was increased, and the cross-questioning of crews who had just arrived was more vital than ever. Happily, we had an accident-free record throughout the war, apart from one plane with engine trouble which was circling over the west coast of Ireland soon after leaving Foynes and jettisoning fuel to facilitate a safe return, when it "grazed" the top of a 3,000-foot high hill on the Brandon peninsula in County Kerry and the passengers rolled out on the heather. Doubtless the lower speeds of the aircraft of those days, little over 100 knots, provided a safety factor in this emergency. We evidently built up a reputation of trustworthiness that came to good in another incident during one winter flight late in 1943, when a flight eastbound from Bermuda for Lisbon came under our control and we promptly recommended return to Bermuda. This action was taken, because we saw that the easterly winds on that route that night would have made completion of the journey to Lisbon impossible. The aircraft turned back and alighted safely in Bermuda, and we later received a letter of profound thanks and commendation from British Airways.

The first winter of the war, indeed the first seven months, were the period of the "phoney war", when the shock of the Hitler-Stalin pact, swiftly followed by the German and Soviet armies' carve-up of Poland, was followed by months in which nothing much happened on the war fronts. That winter was very severe in Europe and produced some memorable weather for south-west Ireland also. Odd cold months during the nineteen-thirties and Britain's snowy Christmas of 1938 had given hints that Europe's more than forty years long period of great predominance of mild winters might be breaking down. And indeed the winters of 1939-40, 1940-41 and 1941-42 were so severe over most of Europe as to cause nearly all war-like activities to be suspended. The first of these winters extended with little moderation to the extreme west of Ireland. With a week-long spell of freezing fog over Christmas 1939 the scenery at Foynes became quite reminiscent of central Norway nine months earlier. Tinsel-like sparkling white ice crystals grew an inch or more long on all the windward-side branches of the pine trees. And, in clear sunny weather in early January, the ice was so thick on some farm ponds that one could drive cars across it while people skated who had not had the chance for many years. On Christmas Day, with a party of British Airways ground staff and, I think two of their captains, I spent the afternoon climbing to the top of the little rugged hill, (Knockpatrick 572 feet) that rises straight above Foynes village. First, we walked through the fairy-tale wood, all hung with glittering ice-crystals, and up till we emerged from the wood and soon from the fog, with the sunshine, and under-foot the grass and ground showing, and around us an unlimited view as clear, and the sky above as deep blue, as on the Alpine summits in similar weather. One had to conclude that, in the calm conditions then prevailing, air that had subsided from greater heights in the anticyclone had reached right down to the hill-top and to other, distant hills north and east of the area, which stood out more clearly than I had ever seen them.

Next day, it fell to my lot to volunteer to drive a young lady, a Church of Ireland parson's daughter, who had been visiting her friends, the Littles, back to her home in Limerick city, 23 miles away, and who was marooned in Foynes by the continued freezing fog. But my windscreen heater, of the primitive type then available - a single, light electrical bar heater held by suction grips to the screen - soon proved unequal to the task, with the prevailing temperature over the low ground well below the freezing point. So I had to complete the 46-mile drive holding the door of the little car slightly open and my head leaning out just enough to see the way ahead without looking through glass. A bitter cold experience, but happily leading to no bad effects.

The life of the airport in a neutral country whose population was known to contain restive elements and sympathisers with both sides in the war was not without its complications. The passengers passing through Foynes included leading members of the allied governments as well as such personalities as Queen Wilhelmina of the Netherlands and others. One bewildered old American lady, clutching firmly the arm of the British Airways traffic officer, who was leading a group along the quay to the restaurant and waiting rooms in the old rambling hotel building which provided all the airport offices, anxiously asked him "You will help me here with the Portuguese, won't you?" All our radio messages about weather and traffic movements were received in code, enciphered by the use of one-time pads in which the numbers of the peace-time international weather codes were changed by the addition of figures printed in the one-time-only pad, which was replaced by a new pad each day. The building was guarded by Irish Army soldiers, and everyone who approached was challenged, even a little group of nuns who were astounded at not being let straight in to collect for a Catholic charity. Only one invading party succeeded in getting in, namely a swarm of bees that settled on the doorway into the forecasting room and delayed a pre-flight briefing - and delayed the flight accordingly - for an hour until a skilled person was found to clear the invaders!

Our staff were very successfully picked for faithful observance of the secrecy of the flight messages. The only times when I came to know the names of any expected prominent passengers (which was none of my business, but it happened a number of times) I always heard the information in my lodgings that I had moved to in the village street, where the landlord was an ex-Royal Irish Constabulary policeman from the days of British rule. He apparently picked the information up from air crew chattering in the bars, either in Foynes or Adare.

The second war winter was again very severe in continental Europe and in Britain too, with frozen rivers, but in the west of Ireland we had seemingly endless slashing hail, rain and sleet showers with winds off the Atlantic and flooding and puddles everywhere. The young Irish soldiers on guard duty at the airport in Foynes village (population 350) challenged me every day as I walked down the hill from the Hotel Ardenoir at 6.30 a.m. past the harbour to the airport building to go on duty. One cold, wet, windy December morning, when the voice behind the muzzle of the loaded rifle pointing at me called, as usual "Halt, who goes there", his words were immediately followed, in anxious tone, shouting "Mind the waater, Sirr" as I narrowly avoided a deep puddle. And during the following spring or summer, when the flights were getting busier, we had another kind of alarming incident. The water supply had failed, as it often did in the summer time, and from that time on two white pails, one of water and the other of paraffin for the Primus stove, each evening were left in the washroom opposite the forecasting room for the night-duty staff to make their tea. Unfortunately, although one pail was an enamel one and the other of some material more like china, the contents of both looked alike. And luckily one morning about five thirty, when I happened to pass the room with the stove on, heating the kettle, I noticed that the soot on the outside of the old kettle seemed to be taking fire rather surprisingly. Quite unusual, I thought sleepily, till I realised what the explanation must be. The assistant must have accidentally filled the kettle with paraffin and put it on the Primus stove to boil! It was just as well we caught the situation before anything drastic happened. But we went without our tea that morning, and we never used that kettle again!

In due course, some time in 1940, with Peters gone back to England, I was formally transferred to the Irish Meteorological Service and received the news by telephone from my new boss, A.H. Nagle, in Dublin that, so far from having any opportunity to be a member of the fine 2-year training course that he had planned for his Irish graduates coming into the Meteorological Service, I would have to be the teacher of the course! I protested at once that I simply did not know enough meteorology to teach a training course four times as long as I myself had received. Nagle was reassuring. That would not really be a difficulty, because he would provide me with a splendid, brand new text book covering all that was required. It was essentially no less than the lecture course given by the Swedish member, Tor Bergeron, of the famous Bergen school of meteorology under old V. Bjerknes. The only real difficulty was that, as Bergeron was too lazy or too much of a procrastinator to write the book of his lectures himself, it had been written by a Czech student of his courses in Russia, G. Swoboda. And it was in German. I knew well enough from this that it would be a very fine book. But my brief school course in so-called "scientific German", actually using Grimm's fairy tales as the set book, was hardly an adequate grounding for this task. I was stunned. But I clearly had to learn enough German and meteorology to keep just ahead of the class. The shock of it started me off on my lifelong career of really serious sleeplessness. I was so nervous that my mind became unduly prone to go blank in the middle of a lecture - about what I meant to say next. The class of seven young graduates were very tolerant towards this failing and in no way took any advantage of me. In fact, I had nothing but good, and often really friendly, relations with them all. There were, of course, minor cultural differences of outlook. One of the young Irish forecasters needs must borrow a bicycle as soon as he received his first pay cheque and ride it into the nearby town to put it all on a horse running that day. He was, sadly, the one who died just a few years later - I believe it was something to do with blood pressure. There was just one difficult moment with him, and maybe it was more instructive than any real difficulty: when it was rumoured that Churchill was thinking of putting troops into Ireland to forestall an expected German invasion, Barney puffed out his chest and said "By God, if they do, we'll fight them both".

For relaxation at Foynes, I bought a little second-hand sailing boat with new red sails. But, before the petrol ration ceased, I also went once or twice to the Kerry hills, eighty miles away. Among them, Carrantuohill (3,414 feet/1,040 m.) among Macgillycuddy's Reeks, is the highest hill in Ireland. I was up there at least twice, when the flying seasons had ended. Once, in October, I was climbing in hill fog with some British Airways friends from Foynes when we got a shock from the unmistakable noise of falling stones on the scree slope higher up, coming quite near us. When the lone walker appeared, he turned out to be a youngish countryman, probably a Kerry farmer, who was crossing the hills with dancing shoes in his pocket to go to a dance in Killarney. On such occasions, climbing in the fog, we put down little piles of stones, as miniature cairns, just 30 to 50 yards apart, to make sure one could find the same way back. The next time I went up that hill, there was no fog. I was alone, and there was such a strong southerly gale blowing that I was repeatedly blown off my feet sideways against a rocky ridge near the path and came down afterwards covered with bleeding grazes of the skin.

The first sailing boat I had at Foynes had been awkwardly big, so that I needed one, or better two, more people as crew to go out with me and manage both sails and tiller. Hence the opportunities for using the boat were rare. One of the few times I used it, we ended up becalmed at the next village down the river and had to climb out and walk home, leaving it to another time-consuming expedition to fetch the boat. Next after that I bought a little canvas canoe from a British meteorological assistant who was recalled to England. I had many more exploits with it, including four times when the canoe sank, which it was all too liable to do, because the canvas was heavy and the little craft was unstable without much keel. The greatest danger arose from the quick changes of roughness of the river Shannon when the tidal current changed. Nevertheless, I used the canoe a lot, and saw a lot with it, exploring bays and islets mostly in shallow waters, both on the Shannon and at the edge of the Atlantic near Valentia Observatory, where there was a little fjord and a considerable island. However, in the light of certain gruesome accidents in which bathers have been taken unawares and drowned by a "freak wave" on our Atlantic coasts in recent years, it may well be that the risks in venturing with such a small craft as a canvas canoe on such a coast are much greater than I realized and perhaps unjustifiable. I have seen an analysis of a swell situation on the coast of Cornwall, where the confusion of crossing waves was found to be due to a swell from a storm off Cape Horn and the other element from a storm near the North American coast off New England.

My most successful sailing boat was the little one with the red sails which I finally had on the Shannon river, which was small enough to be more manageable and not so small as to be often hard to manage. With these small boats it was fascinating learning to identify, and use, the really light "sea breezes" that were liable to blow towards each of the opposite shores of the two miles wide river, ruffling the water near each shore while the middle of the great river could retain a glassy-smooth calm surface.

I also spent seven months, right through the 1940 summer, training the first batch of assistants to make the routine weather observations for the young Irish Meteorological Service. One day while the class was training the telescope of their theodolites to follow the course of a hydrogen-filled balloon in a bright blue sky to measure the upper winds, there suddenly appeared a second balloon in the field of vision. I was called to sort out the confusion as the young assistants needed to know which was the right object to follow. I looked myself and soon realized that of the two objects, which appeared to be of about the same size, one looked nearly stationary and, although the time was 11 a.m., must be an astronomical object - the planet Venus. (I had not heard before that this could happen, but there was no doubt in the end about this solution.)

There were various rumours and alarms that summer about possible invading parties. Some of the country people were more alarmed than others, and I even met an old, old farmer on a back road through the mountains in Kerry, near the Gap of Dunloe, who spoke with anguish in his face about there "being a war on somewhere". It was towards the end of summer in 1940 that I seized a rare chance one fine Saturday evening to take my little open car, the light grey Flying Standard, with the hood down, some miles farther round the coast than I had been before, towards County Cork, and seeing an attractive little glen I drove up the narrow glen road to where it ended at the foot of a hill. Then, taking my camping gear, I went up the steep slope in the evening sunshine and pitched my little haystack type tent where there was a good view at the top of the field. On the Sunday morning, about 9 o'clock, I was boiling my kettle on my tiny methylated spirit stove to make a cup of tea when I noticed a figure in dark blue uniform coming slowly up the steep field and mopping his brow as he came. When he arrived I saw it was a policeman, the local Gardai Siochana. He sat down on the grass beside me, remarking what a hot morning it was, and looked like being a splendid day. I offered to make him a cup of tea, which he gladly accepted. Then, ever so gently, he worked the conversation round to who I was. And he told me that some folk down the glen, who had seen me arrive the night before, had been afraid I might be a German, come to make contact with a Nazi parachutist. So I reassured him, and identified myself as a member of the Meteorological Service and apologised for being the cause of his having to come out so far on the Sunday morning and toil up the steep field. He then thanked me for the tea and went off, evidently satisfied.

There was a lively social life in the little coastal town of Cahirsiveen, near the observatory and quite near the far south-western tip of Ireland, in longitude 10° West of Greenwich. Among the activities were occasional all night dances, which I observed were all very well if you were to be out in the fields next day but not so good if one had a normal day's work indoors to do. There were even open air dances in the summertime at certain cross-roads in the county Kerry countryside. At some there was a special wooden platform for the dancing, but at others just a rectangular concrete slab.

The local garage owner in the town, Fred Mullins, and his wife Nora, who was a very good cook, in whose cheerful white-painted house I lodged, in common with a couple of the young meteorological assistants in my class, were very friendly hosts. There was regularly much interesting and entertaining conversation in their kitchen each evening till a late hour. Neighbours would call, to join in, even in the late evening, once as late as 12.15 a.m. There was also play with their lively, white, rough-haired terrier dog, who could be swung round the well-filled kitchen at the end of a broad ribbon of sea-weed which he held firmly in his teeth. Another type of sea-weed, growing thereabouts, carrigeen, was popularly used to make a light and healthful, mild jelly.

The Mullins's had suffered earlier that year from the dominance of the country life about Cahirsiveen and its ways. He had just set up the first petrol pumps at his garage there and, as bad luck would have it, the date was that of one of the cattle markets that were held along the main street of the town. Before the opening time of his garage, a cow had already been tethered to each of the new pumps and one of them had had to be stopped going away down the street with the tube of the pump still held in the rope that had tethered the beast. In the remonstration that followed, the sad garage owner was told: "We were here long before you, and we'll still be here long after you've gone". Another sign of the isolation of the place so near the end of the Kerry peninsula was that there seemed to be only four or five different surnames among the families living in the little town of 1,500 inhabitants, presumably hinting at long inter-marriage. The O'Connors and O'Neils were probably the commonest. To cope with the difficulty of too many different people in the same place having the same names, a system had grown up of labelling the different Nora O'Neils as "Nora A"

(Pronounced "Nora Ah") and "Nora B" and so on. But if inter-marriage was the cause, it was nevertheless on the whole a very good-looking population, often with quite striking features. The area was, and probably still is too, a refuge for some of the species of wild life that have been dwindling elsewhere in these islands, and in Europe, as their habitats have been ruined by modern agricultural methods and the growth of the human population. One night my sleeplessness could certainly be blamed on a corncrake in the observatory grounds which never stopped his rasping call all night.

Another survival - at least so it was said - in that area was the Gaelic language. The red-haired older assistant, who was by that time the official in permanent charge of the Valentia observatory building was an enthusiast for the Gaeltacht, and he persuaded me that I should drive us both round Tralee Bay to visit a remote spot at the far end of the Dingle peninsula which sticks out into the Atlantic, where the old Irish language was still dominant. (He himself had learnt it, I understood, during his time at a monastery in Fort Augustus in Scotland.) So one beautifully sunny Sunday we duly went down the peninsula by car, driving over seventy miles to an old, partly broken-down quay, beyond Brandon, where a few seasoned boatmen in blue sailing jerseys were out sitting in the sun. As soon as we stopped, my enthusiastic friend at once leapt out of the car and excitedly addressed the men in a torrent of Gaelic. But something was clearly amiss. Possibly the sunshine had made the old salts sleepy. But they had to say - in English - they did not know what my campaigning visitor was talking about.

That summer of 1940 saw such an exceptional prolonged drought - with seemingly endless clear skies - in the west of Ireland that the extensive peat bogs dried out so much that one of them on a peninsula jutting out into the Atlantic near Valentia Island took fire and could be seen smoking away all summer and for months after that, more than half-way through the following winter. I had some pleasant recreation in odd periods after work during those sunny months, no doubt rather riskily, exploring various inlets with the small canvas canoe which I had bought from a British meteorological assistant who was recalled to England. The canvas made the little craft heavier than water, and it sank several times in the shallows, when the water suddenly became choppy as the waves increased in size when the tidal current turned. Once, at Foynes, I had a real fright in that way on the Shannon river near Foynes Island. My canoe was never very stable when sailing, despite its centre board, and it capsized under me when the tide began ebbing fast. To make matters worse, my legs were somewhat tangled in the ropes, but luckily I was not far from the island shore, and I managed to swim there. It was galling that a much bigger yacht was sailing by at that very moment - a rare event in those waters at that time - only a

To Ireland and the New Trans-Atlantic Air Route

few feet away from me. But although I called out nobody heard me nor saw my predicament. So it was left to me, when I had rested sufficiently, to tip the water out of my canoe, reorganise the rope, and get myself back to the mainland at Foynes. My friends among the local population begged me to give up sailing the canoe after that, and the experience certainly put me in a mood of soul searching. It came to me very strongly that after escaping such imminent disaster, it was up to me to use my life in ways that helped other people, and particularly the saving of lives, through my job of concern with the safety of the trans-Atlantic flying.

It was while I was at Cahirsiveen that Norway and Denmark were invaded by the Nazis. I was horrified and very deeply shocked - as was the civilized world - that two countries so uninvolved, and innocent of any great power-like machinations of their own, should be thus over-run by the tide of war and, indeed, with the use of a horribly deceitful stratagem whereby many members of the invading force that morning of 9th April 1940 emerged with their weapons from the holds of what had seemed to be ordinary trading ships in the ports of Norway and Denmark. Of course, we had already seen the carve-up of Poland that swiftly followed the surprise of the cynical Nazi-Soviet pact. And there was more to follow. Clearly I had to re-think my pacifism in the new circumstances of an increasingly savage world. In Foynes, there was from that day tangible evidence of the changed circumstances in the outside world. Two Danish merchant ships of the Maersk Line had sailed into the Shannon to avoid the risk of capture by the Nazis. From that date on, throughout the rest of the war, I took the Norwegian government-in-exile's weekly newspaper, *Norsk Tidend*, published in London, to get the news smuggled out of Norway of conditions in the country and affecting the Norwegian people everywhere in the world. With all the British meteorological staff apart from myself and Paul Brown withdrawn from Foynes, I, with Paul became solely responsible for the Allied secret codes. There were also two refugee scientists who had joined the Irish Meteorological Service at Foynes, Professor L.W. Pollak from Prague and Dr. Mariano Doporto of Santander, who had been the chief Spanish meteorologist on the Republican side. They were both, rightly, considered entirely reliable. I shared an office for some months with each of them in turn, first with Pollak and then with Doporto, and became good friends with both, although they did not get on so well with each other. Their backgrounds and traditions were too different.

Pollak, the typical learned Jewish professor, kept up the very formal manners of central Europe. Each morning on arrival, and again before and after lunch, and at the end of the working day, he shook me by the hand and said "Good morning (or good afternoon), dear Kollega". He took out his pocket watch and placed it on the desk in front of him as long as he was there and gathered it up before leaving. We had a good deal of friendly converse about science and the methods in use, about politics and the war, and about people in all the countries of our acquaintance. Going away for his lunch, or for the evening, he would repeat "Goodbye, dear Kollega. I wish you a good meal (or a good evening)". Such manners gave a strange and unusual impression in the wilds of far away County Limerick. Pollak, in due course, left us to go to Dublin to set up an Irish Geophysical Institute with a small establishment, for which he had Eamon De Valera's special blessing.

Doporto was a typical Spaniard whose dark eyes sparkled with fun, but who was equally committed to meteorology and the Met. Service job. In due course he, too, disappeared to Dublin, sharing an office for some time with Nagle, the director, before himself becoming director when in 1945 Nagle went off to a post in the United States Weather Bureau. During the months when he and Nagle shared an office in Dublin a slightly dangerous situation developed, in which they developed a common mind that was somewhat at odds with the staff of the Service out around the country, and notably reinforced their entrenched position, assuring themselves and each other how right they were, while the pleas from the out-stations that the Service was taking on too many commitments went rigidly unheeded.

Before coming to Ireland, when the Republican government cause had collapsed in Spain, under the onslaught of Franco's insurgents, supported by Hitler, Doporto had become an exile and made his way rather quickly across France and Britain to the post he had been promised in Ireland. His wife and two young boys, aged 7 and 9, were already exiles in a camp in Tunisia, and it took a month or longer before they could rejoin their papa in Foynes. They could speak only Spanish and French, although their father's English was good. The boys had to go to the local school in Foynes where, as in all the National Schools in the Irish Republic, all subjects were taught in the Irish language - Gaelic! That must have been a severe test, complicated further by the fact that, as soon as they were out of the classroom, all the children spoke English - another strange language for the Doporto boys. Yet it was only three weeks before the older brother put up his hand in the mathematics class to ask a question, necessarily in Irish! Such is the prerogative of the childhood years!

Doporto was in charge at Foynes during most of 1941, but when he left to join Nagle in Dublin it fell to me to take charge of the Trans-Atlantic forecasting service. To this end I had worked out a system for drawing maps that indicated the winds blowing at about 10,000 feet (three thousand metres) height over the Atlantic, this being roughly the height at which the

To Ireland and the New Trans-Atlantic Air Route

passenger aircraft of those days flew. The system depended on a series of rules of thumb about the probable changes of air temperature - and the changes of density implied thereby - with increase of height over the ocean. Those rules were based on the normal ocean surface temperature in any part of the ocean combined with consideration of where the wind had come from. The system was explained in all necessary detail in a Technical Note published by the Irish Meteorological Service. It was put into use immediately and served us well throughout the remaining years of the war, while there was no other relevant information available to us.

In August and September 1942 I went with Nagle and three other representatives concerned with air traffic control and radio communications from the Department of Industry and Commerce, Dublin to Toronto to take part in the first Trans-Atlantic Air Services Safety Organization conference. After that, we went on to New York to see the facilities which the Americans were using. Our westbound crossing took about fifteen hours, as was normal at that time, on the shortest route from Foynes, near Ireland's west coast, to Botwood on the north coast of Newfoundland. In spite of the long hours we spent in the air, that flight was luxury indeed in comparison with later times. The flying boat was furnished with a few small round tables, each with four soft armchairs round it. Each passenger also had a bunk aft to sleep in. But I did not attempt to sleep. My first Atlantic crossing was to me a wonderful chance to see the cloud development and patterns over the ocean, as they really were. We also saw one or two small icebergs in the water on the last stretch before reaching Newfoundland. At Botwood, in the Bay of Exploits, the sight of the prevailing vegetation of birches and pines, and heath plants, was a glorious reminder of Norway and made me realise how much I was pining for that.

Once ashore, we had to travel to the airfield some distance inland, at Gander, in an upland region of the interior beside the long Gander Lake. To get there, we had to travel some thirty miles through the forest on the single track of the main Newfoundland railway, which connects the west and east coasts. We rode on a heavy iron-built trolley, like those used by workers on the line in other countries. At each station we came to - there were not many - we had to stop and ask the staff for news of any train on the line ahead, and when one was expected we had to lift our trolley - "everyone together, heave!" - off the track and wait for the train to pass. Even so, every time we came to a bend, one looked ahead and listened anxiously. The Royal Air Force, as owners of the trolley, knew what it was about and insisted before we travelled that we each signed a "blood chit", promising not to hold His Majesty King George VI liable for our deaths or for any injury in case of accident! But all went well, and we duly reached Toronto and the conference. The worst that befell us was when we were entertained at a reception by the Canadian government in the Royal York Hotel, where twenty-four kilted pipers marched into the upper room in which all the guests were assembled, piping lustily. Happily, I believe no-one's eardrums were injured! The conference was an undoubted success and set up procedures that provided the Trans-Atlantic air route with sound working practices.

All that I remember of the New York meetings that we went on to is the hot sunny weather which greeted our first few days. When one emerged from the air-conditioned airport offices building, one involuntarily supposed at first that somebody had left the heating on! But after very few days a deliciously cool north-westerly breeze swept the avenues of Manhattan under a marvellously clear blue sky. There were some entertaining, and a few embarrassing, moments when the Irish party was introduced to night clubs after the day's work was done. Scantily clad young dance hostesses came to the table and asked one to dance. But a well-built sailor, probably Scandinavian, in about his thirties, who evidently felt even sleepier than I did after the long hours of intense business discussions, just gave up and put his head down on his hands about 3.30 a.m. and was instantly asleep. I was really quite happy when, as youngest member of the Irish government party, I was detailed to go back to Ireland ahead of the others.

However, my journey was not destined to be particularly swift. I got no farther than Canada's Maritime Provinces - to be specific, the airport at Moncton, New Brunswick - before the weather took a hand in frustrating my intended progress. All the airports ahead in Newfoundland were reported blanketed in fog, preventing any flights, and so they remained for the next four days. Where I was, however, in New Brunswick, and in Nova Scotia, we were blessed with totally clear, cloudless skies for all that time and with the lovely cool west to north-west breeze that I had just sampled in New York. So I hired a push-bicycle, and spent those days happily exploring the countryside and the little places with sparkling white wooden houses, and a jumble of French, Scottish and English names. They all looked beautifully clean and quiet and wholesome in the bright calm of the autumn sunshine.

In the end, of course, but only after I had benefited from a very refreshing break, the weather in Newfoundland cleared and I was flown to Gander airport and made my way on to Botwood by a repeat of the bizarre rail journey. Before long I was airborne again, bound for Ireland, in a Pan American Airways Clipper. But when we were only one and a half hours out over the Atlantic, the captain who already knew me from Foynes, came down from the flight deck to the passenger section, tapped me on the shoulder, and asked if I would come to the upper deck with him. We were scarcely there before he fished a bunch of telegraph messages on pink paper out of his side pocket, thrust them into my hand and said: "Take a load of that. What do we do?" It was undoubtedly an awkward situation. Our arrival in Ireland was now expected to coincide with that of a slow-moving cold front which was so aligned (from south-west to north-east) that it would probably blot out Foynes, and simultaneously all the alternative airports which we could reach in Britain and Ireland, with fog and very low cloud. And the warm, very moist air ahead of it could be expected to blanket similarly all or most of western and southern England as well. Once I had digested this, the skipper added: "What they don't know is that everywhere behind us has closed in with fog too". Things were all the more awkward for me, being asked for this emergency advice, because my month-long engagement with the administrative conference and discussions had given me no opportunities to see weather maps and keep in touch with the current situation and how it was developing, apart from the few minutes when I had rushed up to the forecast office at Gander airport before leaving.

However, after a minute or two's thought, I realised that I could say something practical on the basis of familiarity with the Shannon river and the particular terrain within a radius of fifty miles or more surrounding the airport at Foynes.

In that part of Ireland the extensive lowland landscape of Counties Limerick and Clare, drained by the wide Shannon river flowing west through the middle of it, is broken by a sparse scattering of mostly rather rugged ridges of hills at various angles. Some of the lower ridges, rising to only 300 to 650 feet are quite near the river, but others rise to 1,000 to over 2,000 feet, particularly east and north of Limerick itself. Some of the low ground is quite rocky, but some miles east of Foynes the river Shannon opens out into a watery expanse four, five or even six miles wide. Much of this water is quite shallow, but there is a mile-wide central channel that has more than five fathoms depth, even for eight miles farther inland than Foynes. The effect of this geography is that I was able to tell the captain that, although very low cloud was common with slow-moving cold fronts, from the north-west, I had never seen an unbroken extent of very low cloud below the heights of the hills over that area. Always there seemed to be rifts in the cloud cover torn by the hills, though there would be other spots where the light breeze caused the low cloud and fog to bank up against any windward slope. In the case of a widespread night radiation fog, in otherwise clear winter weather, openings would appear over the river owing to the relative warmth of the water. I told Captain Cone that I believed he could depend on finding extensive enough areas clear of the lowest cloud somewhere within ten miles up-river from Foynes. And that is exactly what we found. The Clipper alighted on the water in the wide area where the rivers Fergus and Shannon joined, and then taxied eight to ten miles down river to the mooring in Foynes harbour.

It was a neat example of the virtue of float planes for exploiting alternative landing possibilities. There was another case, in the first or second year after the war ended, when a Trans-Atlantic flying boat came down on the ocean surface south of Newfoundland and stayed afloat for rather over twenty-four hours, allowing time for everyone onboard to be rescued before it finally sank.

Keeping the weather service arrangements working, of course, required frequent consultations between Foynes and the director in Dublin, 130 miles to the east, on the other side of Ireland. Mostly, telephoning sufficed, though it was a wearisome business in those days: one had to wait, as patiently as one could, to get the line linked up from one local switchboard to the next, right across the country. And, almost always, voices would cut in from one switchboard or another to ask "Have you finished?" before the line had ever been linked right through. At other times one or other of the switchboards along the line would simply disconnect one on the assumption that the call was finished before the call had been successfully linked all the way through. This was such a regularly harrowing experience that I have never liked telephoning since, and I am always mildly amazed when a call is completed and turns out to have been unquestionably useful. But Nagle, the director, was a great user of the telephone. Some of his calls went on for an hour and a half. Once or twice a year it became necessary for us to meet in person. And after the petrol ration was finally withdrawn, that meant going by train. There were not many trains. Several times I made the journey through the night with the milk train, which kept running twice a week. With its leisurely progress and slow operations, at the stops where loading took place, I found it possible to get enough sleep, sitting up. Nagle, whose preferred hours of work were from 11 a.m. or midday to midnight used to accuse me of looking like a sleepy owl in the late evenings when he was still putting his arguments to me. Finally he would give up and let me make my way back to my hotel.

Once, in late November, when there was no more Trans-Atlantic flying for that year, I cycled all the way across the country from the village of Askeaton, Co. Limerick, where I was then lodging, the one hundred and twenty miles to Dublin for a brief cultural break. I chose a bright, sunny day when there was a strong westerly gale blowing to speed me on my way and completed the journey in just ten hours from 11 a.m. to 9 p.m. The winds in Ireland were so predominantly westerly that I was apprehensive about the

To Ireland and the New Trans-Atlantic Air Route

likely conditions for my return westbound. I arranged to spend a night on the way at Naas. But when I set out, after the weekend, a gentle breeze began to pick up from the north-east. And, although it was never strong, it continued to help me along all the way. I could scarcely believe my luck.

Journeys were occasionally necessary either for shopping in Limerick, 23 miles east of Foynes, or for recreation. They usually turned out to be somewhat of an adventure, whether made by bike or on the then twice-a-week train service. One happy feature of Irish life made things easier. It was almost impossible to be too late for the bank or for a shop. Provided one's arrival was not very late, some kind person could nearly always be found who would answer one's pleas from inside the door and open up to do whatever was needed. Many of the features, such as this, of Irish life which have been thought ridiculously unruly or laughable by people from other countries, especially from England, are really customs of value to rural life and in scattered communities anywhere. Parallels could be found, for instance, in Norway or the Alps where the bus or its forerunner, the horse-driven "diligence", would not depart without any passengers off the fjord boat or the train who might be intending to join it. In the early days of the airport at Foynes, when the first British Airways catering officer, Sidney Cook, a Londoner, arrived to take up his duties, he soon became friendly with various people in the village and one day, when he had arranged to go with one of them on the 10 a.m. train from its terminus at Foynes to Limerick, he somewhat doubtfully agreed at 9.45 a.m. to have a "jar" of Guinness in the Shannon Hotel bar, 100 yards along the street. He was taken aback when pint glasses were brought. The train was in the station, standing ready, with steam up. Ten o'clock came and Sidney's companion showed no sign of worry. He said he had had a word with the engine driver, and he "wouldn't go without us". But at ten past ten, when the glasses were still about half full, the engine driver himself appeared and stuck his head round the door and said: "If you two don't come quick, I'm off".

The supremacy of the old country ways was demonstrated to me and one or two friends from British Airways at the Shannon Airport whom I was driving to the Kerry hills for a day off in the autumn of 1939, when we came to the town centre of Kilmallock where a cattle fair was in progress. As usual on such occasions, many of the prudent shopkeepers had boarded up their windows and doors. The street was thronged with farmers haggling and gossiping, and their animals standing in the midst of them, while along the kerbs on both sides the little donkey carts were drawn up. Ours was almost the only motor vehicle in sight. But, in any case, the line of sight did not stretch far in any direction before it was blocked by people, carts or buildings. As we nosed our way very slowly on gingerly through the crowd, a little clear space opened up briefly near the middle of the street. At that, a cow four or five paces ahead that had just seen us took fright and leapt over the nearest donkey and its cart to our right. But the leap did not quite clear the poor little beast, and the cow came down cross-wise astride the donkey, which then sagged to the ground. Several hands came quickly to the rescue, while the cow pushed herself up and wandered on, unaided, into the open door of the nearest shop. We last saw her, as we crept by in the car, surveying the confused scene outside on the street through the shop window, while she placidly chewed the cud.

The British Airways Traffic Officer at Foynes, Hilary Watson, a tall thin young man, half Scots and half Hungarian nobility by birth, was a memorable character. He had been moved from a post in India for the new Trans-Atlantic air service. He had a figure somewhat like a telegraph pole, but was fond of wearing his green Buchanan tartan kilt. His Hungarian parents had endowed him with twenty-one Christian names, nineteen of them German. So, after consulting his airways employers, he considered it advisable in war time to change his name by deed poll and get a new passport before the expected hostilities between Britain and Germany broke out. On arrival in Ireland, he already suffered some difficulties and frustration because, although some months had passed, his new passport had not reached him. When it finally came with his names given as "Hilary Buchanan Watson", he noticed an asterisk which referred the reader to the last page of the passport and there he read: "Hilary Buchanan Watson formerly known as "Karl Emil Ludwig .... Hilary Buchanan Watson". In spite of this, I never heard that any embarrassment ever arose.

There are, of course, aspects of the historical relationships between the British and Irish peoples that both sides find difficult to deal with. For centuries after the forces of an English medieval king, which had been invited into Ireland about AD 1170 to help decide an internal struggle, were not withdrawn at the end of that mission, England was regarded by some part of the Irish population, not altogether without reason, as an occupying power. And I myself sensed that one or two of the twenty strong Irish meteorological staff at Foynes airport took up to two years to discover (and convince themselves) that I, as a Quaker, did not share quite the usual British imperialistic attitude to Irish questions. One of these old Irish colleagues of mine has since written about the tragic history of the lethal Irish potato famine in the 1840s with great appreciation of the helpful part played by the Quakers of that time and that the British government in London at that time did all it could to alleviate the suffering, once it understood the realities of the situation. The potato blight affected many areas of Europe in those years and damaged the crop, but Ireland was more

To Ireland and the New Trans-Atlantic Air Route

dependent on the potato crop than anywhere else, because it was the only crop which could produce the actual bulk of food needed to fill the stomachs of the big Irish families on farms often of only 3 acres (little over 1 hectare). It was a tragedy waiting to happen, but was made all the worse by the fact that the potato had never been known to fail before. It was a new disease, brought to Europe in a ship-load of South American potatoes. It struck at a crop that had been regarded as the "bulwark" of rural life in Ireland. The only possible solution was mass emigration and within six years the population of Ireland was halved.

Once or twice in my Spanish colleague, Doporto's reign, before I became too heavily occupied by the increasing air traffic across the ocean from Foynes, I was transferred for a week or ten days to run courses for pilots of the Irish Army Air Corps at their air base, Baldonnel, near Dublin. At their invitation, I went up with them for an occasional sample flight in two-seater open aircraft identical to those I had flown in with the RAF at Montrose. Again I took a few sample upper air temperatures and encountered just the same problems with them as with the young RAF trainee pilots. It was hardly possible to persuade them to level out the flight for long enough to allow the thermometer to settle down to a constant reading. But one of them did give me a very memorable tour in roughly one hour on a brilliantly sunny afternoon. We flew southward through broad valleys, passing the Wicklow Hills and Kilkenny to look at the rock of Cashel which rises abruptly out of the Tipperary plain and is topped by striking medieval and later ecclesiastical buildings and remains. From there we flew on to the Galtee mountains which rise almost astonishingly between the lowlands of counties Tipperary and Limerick to reach at one point over 3, 000 feet. We then flew back to Dublin more or less direct by another route, somewhat farther west and north. It was a splendid tour on a wonderfully clear day.

I came to know one or two of these young pilots quite well during visits to the Army Air Corps and even received - a nice surprise for me - punctilious repayment of a small loan that I had made to one of them. To my shame, I had misjudged him, imagining that he was wilder than he was and likely to be irresponsible in petty financial matters. This lad, like his colleagues, was so keen to fly than he found serving in the Irish Air Corps extremely frustrating owing to its restricted operations as a neutral in the war. Already, one dark night in 1941, he had quietly gone off with one of the Irish machines to join the RAF. But he was intercepted and bought back. In 1944, he tried the same again, but this time, being evidently surprisingly unaware of the course of the war, his object was to join the German Luftwaffe. But his father had good friends in the RAF and was able to arrange for him to be intercepted again, over Cornwall, where he was forced to land and was brought back home.

Another of the young officers, told a memorable tale one evening over a round of stout in the officers' mess of how he had been feeling a similar frustration in 1936 because of the lack of openings at that time in the Irish Air Corps. He had been pent up about it and had discussed it often with his friends in Cork. One of them had said predictably: "Why don't you go and join the RAF". But to that his immediate reply had been "With my family history I could never do that". (One near relation of his had been a leading light in the Easter Rising in 1916.) "Well then - go away and join Mussolini's air force" was the next suggestion in reply to that. And that is what he did. But it seems the Italians did not trust him fully and had put him on transport duties only. (After all, there is always confusion on the continent about the various countries in the British Isles, and Nazi propaganda about the time we were talking was being posted to supposed sympathisers in "Cork, England".) But he did have some interesting experiences flying for Mussolini's air force. He had flown a number of missions across the North African desert from Tripoli to Abyssinia by way of Kufra oasis. There, on one occasion, he and some Italian colleagues had been feasted by the sheikh, which led to them being confronted by sheeps' eyes on a dish. That caused C???? some embarrassment. But the sheikh had another surprise for him, telling him proudly that there was a girl in his harem who would enjoy meeting him. She turned out to be a red-head from Tipperary. She had arrived in North Africa with a dancing troupe travelling from country to country. The company had gone bankrupt somewhere in Tunisia when the war began, and the performers had been left stranded.

Although I always enjoyed very good relations with the Irish Catholic population whom I came in contact with on the met. service staff and in the lodgings where I stayed, there were undoubtedly some sinister memories still about from the troubles only twenty years earlier, around 1920, when there were numerous murders and occasional mass killings perpetrated by both sides in the war between the IRA and the Black and Tans. Those who had been too close witnesses of those events may not have been able to help hanging on to their gruesome memories and the immediate hatreds and desire for revenge aroused, as I had myself seen at close quarters in Britain and Germany, and having seen the same mind at work in France. This is the most difficult problem that any would-be peace-makers have to digest and try to deal with. I myself was taken one November night by some of the young Irish Army Air Corps Officers from Baldonnel airfield to visit the family living in a grand house somewhere just beyond the Wicklow Hills, where I was shown the carefully kept Irish army tunic with a single bullet hole just near the heart that had been worn by one of their relatives who was shot during those years in the house. One can only try to soothe and sympathise. The one clear lesson is surely that nothing is solved by such actions. Violence breeds only violence.

As I was five and a half years in Ireland, a country I had not known before, and could not contemplate marrying into the Catholic faith - although I could see things to admire in it - the question of whether I could make any real friends, and who my friends would be, was a vital one. I was on very good terms throughout with so many, especially among the meteorological and other airport officers' staff, and for over two years lived in lodgings along the village street in Foynes, which I shared with one of the Irish forecasters and two of the observing assistants. We had a lot of fun and plenty of the amusing conversation, frequently invoking familiarity with the saints, with which Ireland abounds. But on world problems one ran up against limitations. So many of the Irish sympathised with Franco's Spain, although my friend and colleague, Mariano Doporto, the refugee from the republican side in Spain was always treated in a very friendly way. There was one forecaster on the staff who had had a wider, eye-opening experience. Vincent Guerrini had lately returned to Ireland from an apprenticeship, working in one of the optical firm Zeiss's factories in Germany. His view of the supremacy of the priests in Irish life was freely and frankly spoken and showed an insight that none of the other Irish staff voiced: How comfortably they lived now and how they lorded it over the rural populace when a hundred years earlier "there wasn't a priest in Ireland with a shirt to his back".

At the other end of the social spectrum, I certainly was not attracted to the society of the landed gentry, the survivors of the old "Anglo-Irish Ascendancy". Of course, there were exceptions, odd ones whose cultivated individuality and out-of-the-ordinary preferences evoked a certain charm. There was one lone bachelor, by the name of Wardle, still living in what had been his ancestral family home about ten miles from Foynes, who used to come into the Ardenoir private hotel one night each week while I was there, to eat the hotel's rather simple dinner but grandly dressed for the occasion. He invited two of us to dinner once in his own home and had to push back the dusty piles of books on one corner of his polished mahogany table to make a little room for us to eat.

Some people in Ireland with names like Lamb attach importance to spelling their name with an E on the end - a custom which never attracted me, particularly when a snobbish element is attached to it. There are, I think, too many silent letters, particularly final E's, in English words which only complicate the spellings and pronunciation. One red-haired Irish colleague, who through some unknown quirks of his ancestry boasted the Italian name Guerrini, used to say when any shopkeeper asked for his name: "Green - spell it any way you like". Maybe that is the most sensible way of dealing with a pointless controversy. Some of the gentry about Co. Limerick were quite cranky in their addiction to English ways. They liked to show where their loyalty lay during the war by setting their clocks to "Double British Summer Time", two hours ahead of Greenwich Mean Time and an hour ahead of virtually all their neighbours. Similarly, they made a point of hanging a framed portrait of King George VI and his Queen Elizabeth, now the Queen Mother, even if it had to be placed tactfully in the downstairs WC - probably a solution equally acceptable to the most inveterate Irish Nationalist. The situation as regards time keeping was, in fact, even more complicated in the rural areas, since some priests ruled that even Irish Summer Time (an hour ahead of Greenwich) was also not in accord with God's will. People must stick to "God's Time", and that revealed that God kept Greenwich Mean Time. But even that did not solve all the problems. It was clear that anyone concerned with catching a bus or a train must set a clock to Irish Summer Time (= BST!), but the schools had to find a compromise that would not push anyone too far out. They therefore found that "Half Time" (halfway between GMT and BST) was most practical.

In Askeaton, on the main road between Foynes and Limerick, where I lived for the last two years or more of my time in Ireland, the priests' rule was not confined to the question of time-keeping: a piece of note paper on a little notice-board on the river bank laid down in the priest's handwriting the times at which the boys or girls of the place (never the twain might mix!) could bathe in the local river.

Non-Catholics in the county amounted to only one per cent of the population. Consequently the finding of marriage partners was always a difficulty for them, and there were many examples of big age gaps between man and wife. It was therefore never very likely that I would find a wife in that country. In fact, my age fell right between the generations in the families I came to know best. Nevertheless, some of my friends in Ireland, where claims to the "second sight" are even commoner than in the Scottish Highlands, assured me that "some day Miss Right will come along".

Looking back, I have no doubt that the most dangerous moment for me in my bachelor life came - one could say it stole on me unawares, as of course it would, since one can only assess the likely consequences afterwards - on the night of the Japanese attack on Pearl Harbour in the Pacific on 8th December 1941, which so largely destroyed the great American fleet in harbour. That event had its effects, even in a small remote Irish village on the other side of the world. The main flying season for 1941 at Foynes was

To Ireland and the New Trans-Atlantic Air Route

over, and many of the airport staff were at a dance in the new, but unheated, village hall when the news reached us on someone's radio. The night was clear, calm, and severely frosty, with a starry sky. A sense of shock and bewilderment about the future instantly spread through the dancers as news of the size of the destruction became known. This, as it were, electrified us all. Like Ireland, the United States had been a neutral country. But that was all changed now. One wondered where it would all lead to and what would happen next even in that peaceful corner where we were. But there was an air of excitement too, counteracting the tendency we all felt towards apprehension and despair. We were cold and alarmed, and I suppose we unthinkingly held each other tighter and danced closer together. I was dancing at that moment with the only non-Catholic girl in the village, just three or four years younger than I was. She was a girl who some of my young Irish colleagues had urged me to marry, although it did not seem to me that in the long run she had the right background for me. "She would be a nice girl for you", they said. "Why don't you?....." Though she was of mixed Anglo-Irish Protestant and German descent, she had no experience of the world outside southern Ireland. She worked, and lived, in the office building in the village of the local great estate, Mount Trenchard. She may have had some hopes of me, but she certainly also met from time to time most of those who worked in and around the airport offices at Foynes, including Americans who probably interested her too. She seemed a nice, quiet girl. But I, too, had other interests and was extremely busy besides. In the atmosphere of that night, uncertain of my own intentions, I waited for her after the dance on the footpath outside between the village hall and the Estate Office that was her lodging. There were plenty of other folk about, people going home and couples cuddling. In the frosty moonlight, she saw me and said "Hallo" quite boldly. But with background thoughts of what complications any hasty liaison might lead to, I merely said "Hallo" and let the moment pass.

When I think now of the rich happiness that has been mine, indeed ours, ever since I met Moira, my wife, nearly six years later, and somehow won her love, I am endlessly thankful that I avoided the rash but wildly tempting decision in the excited atmosphere of that night in December 1941.

Early in 1943, I moved to more spacious and elegant lodgings on the square in the middle of Askeaton, a small town eight miles east of Foynes on the Limerick road. There were full-length white lace curtains in my upstairs bedroom at the back of the house and even a grand piano, which I was allowed to play (though my standard of playing has always meant that it was strictly for my own peace of mind only). The elderly spinster sisters, Anna and Kathleen Fitzgibbon, who owned the house, were very good to me and even gave me a beautiful blue rug when I finally left Ireland. They were sisters of a family of Catholics who were successful farmers in the parish. But in that area, luckily for me, there was another group of families who were equally independent of the closed society of the "ascendancy" and of the Catholic majority, and with them all I had an easier relationship. These were the "Palatines", descendants of the Protestant refugees from the Palatinate of the Rhine in Germany, who fled before the invading army of the French King Louis, who were settled on land allotted to them in this part of Ireland by the victorious Duke of Marlborough in 1709. Nine generations later many of these folk still maintained their independent Protestant identity as Methodists, though some had married Catholics or drifted to the (Anglican) Church of Ireland. They were often conspicuous as the most methodical and successful farmers in the area. Others were prominent in the professions as bank managers, engineers, doctors and so on. One family, in particular, the Crosses, who lived in the bank house just across the square from the Fitzgibbons were very good to me. They had me most Sundays to Sunday dinner and every Christmas. One year I gave their two boys two model aeroplanes for Christmas, and I have to admit that it took us a year to get one of them completed and able to fly. The wife's brothers, Ernie and Fred Shier(the name had been simplified years ago from Schreier), had two of the best farms in the neighbourhood. It was said that the German bibles these families brought with them from the Palatinate (Rhein-Pfalz) were still in use for the first three generations, but after that inter-marriage mainly with people of British stock had resulted in the language being lost. Their sympathies were both British and Irish, and like the other Protestants in the Irish Republic, they suffered no hostility or adverse discrimination whatever. Living in a peaceful part of the Irish Republic, one could regard the centuries-old Anglo-Irish problem as a problem solved apart from the continuing tragedy of provocations by both sides in the north of Ireland.

My sojourn in Ireland was, however, destined to come to an end in 1944, a little before the end of the war. The reasons had nothing to do with the war nor any Irish question but were once again a serious crisis with my employer, A.H. Nagle. I must be a difficult person. Nagle, as the first Director of the Irish Meteorological Service, was understandably ambitious that his service should play a part in all relevant branches of the nation's life. But, as I saw it, from some time in 1943 or early 1944 he took on, and instructed the main Trans-Atlantic weather forecasting office at Foynes to perform, an increasing range of services to the community without being able to employ and train any new staff. In my view of things at that stage this extra work could not be undertaken without jeopardising the attention given to the safety of life on the Trans-Atlantic flights. So, when matters came to a head in November 1944, I resigned. And, to my very pleasant surprise, without applying, I was promptly offered a job back at the appropriate level in the United Kingdom Meteorological Office. Moreover, I was tactfully placed in the forecasting office at RAF Transport Command Headquarters in Gloucester, a post where my sensitivities about certain military involvements were hardly likely to arise.

CHAPTER 12. INTERLUDE

So, after five and a half years involvement in the Trans-Atlantic Air Route, the time had come for me to leave Ireland. During the statutory four weeks of my resignation notice, I was switched to Dublin Airport (Baldonnel) to a supernumerary post and was invited by the kind Pollaks to stay with them in their Dublin city flat. Professor Pollak, short of stature and, in appearance, very much the central European Jew, had a much taller wife who was also Jewish. Their flat was a charming outpost of Austria or Austro-Hungarian culture. He was by that time well established in his director's chair as head of the small Geophysical Institute, but he retained his contacts with people at all levels in the Meteorological Service and he told me that my stand for caution about the risk to air safety from the new proposals overloading the meteorological service had aroused respect and gratitude from the staff of the service at Foynes.

After that and a quick return visit to Foynes to collect my little car, now once more provided with a petrol ration for the journey and loaded to its canvas roof, I said my good-byes in Co. Limerick and Dublin and drove by way of Dublin to the boat for Holyhead and England. In the Customs Hall at Dun Laoghaire I faced an embarrassment. My good friends in Askeaton had made it impossible to leave without taking gifts of a pound of butter from each of two houses there, the Crosses' and the Fitzgibbons'! The permitted export limit was only half a pound or less for each traveller. I had four times as much. I took some trouble to tell the Irish Customs Clerk about it beforehand, but apparently only embarrassed him by speaking of it. When it came to the inspection of passengers' luggage I immediately turned the clothes in my bag back to reveal the illicit export, but the Customs official equally promptly turned the clothes down again, waved his hand, and marked my bag as passed OK. While this was going on a pound of butter was confiscated from each of the neighbouring passengers' bags on either side of me. They had not declared anything!

At Dundalk, the frontier station on the Dublin-Belfast railway, the wily Irish Customs officers had a simple way of embarrassing the many butter smugglers travelling north. The travellers were all requested to leave the train and were shown to seats in an overheated waiting room with a roaring fire. They only had to wait until the butter itself started melting rapidly.

I was back in England by mid December 1944 and stayed the first few days with my old friend Dennis Conolly on the Surrey Hills near Kenley. We went for daily walks on the hills while I heard about his experiences with the Friends Ambulance Unit in Austria, and exchanged news, against the background noise of V2 rockets crashing on the nearer parts of London. For Christmas I went to my parents in the guest house where they were living in Cambridge. And, at the beginning of January 1945, I started work in the forecasting room of the Meteorological Office at RAF Transport Command in Gloucester. Once again we were having a spell of severe winter weather that had begun in mid December and went on right through January. I was billeted in a terrace house in Gloucester, but it was so cold that, after the evening meals the only practical choices were either to go to bed to keep warm or back to the office, which was well heated. I tried each on different days. That office was a very friendly place, but the forecasting room was too big so that on a busy night one lost a lot of time and energy just in walking about the room to fetch and use papers hanging on clips on the walls. That was a quite serious drawback. After a while I got another billet in a rather smart and comfortable house on the outskirts of Gloucester at the foot of Tuffley Hill, a place that I remember for the congenial and kindly owners, the Butts, and how well I slept - the sleep of exhaustion - despite the tanks exercising up and down the road just outside my bedroom when I had come from night duty. Very tiring nights they were because of the busy flights, mostly to the Mediterranean and beyond, and all the exercise in walking back and forth across the forecast room.

The weather had suddenly turned mild and sunny from the beginning of February and it was a lovely, sunny spring that followed. The charms of the Cotswold countryside and of the Forest of Dean and Herefordshire, which I re-visited by bus from Gloucester, and of all the blossom in the gardens in the city itself, in the continual spring sunshine of 1945, made my return to England a very happy time. In April 1945 there were even some unusually hot days. Soon, with summer coming on and the war ending, it became a generally happy time for most people. I took the train by way of Birmingham to Glasgow and slept the sleep of the just, sitting up in a crowded carriage, and went on to Fort William for a short break in the Highlands. Despite the advanced spring in Gloucester it was still winter there and, rashly, I climbed Ben Nevis alone in the snow, making as sure as I could all the way that I could follow my own footsteps back if need arose. (The weather was quite stable, and I had no trouble.)

Later, I was transferred to Hurn Airport, in the edge of the New Forest, near Bournemouth. It was a delightful place, essentially a clearing among the pines and birches, where one walked along little sandy paths - being careful of the adders sunning themselves - to go from our office in the base of the Control Tower on the edge of the airfield to the officers' mess for lunch. There was a lot of sun that summer. Hurn at that time was doing duty as London's main airport before the new London Airport at Heathrow was opened. Occasionally, I had to welcome visiting foreign meteorologists on behalf of the director in London, and entertain them briefly, before they went on by train. I particularly remember the day when I had to welcome the head of the Indian meteorological service, the Director-General of Observatories, and one of his senior staff, who was accompanying him. I remember being particularly impressed by them. They wore the complete old-style London civil servants' cum scientists' clothing, immaculately dressed, speaking excellent English, and with faultless manners. I found myself looking carefully at them again just to convince myself that they really had black skin. These two gentlemen presented so perfectly the image of the typical London higher civil servant of that period - by their immaculate suits, their polished manners and their speech, that they magnificently demonstrated (if I or anybody else needed such a demonstration) the utter irrelevance of colour. I was fascinated and completely forgot the nervousness I had felt at the outset about doing duty for the director in London welcoming his important foreign guests upon their arrival in Britain.

While there, I was billeted in a rather wealthy house with two elderly spinster ladies on Queens Park Avenue in Bournemouth. On Christmas Day 1945 I committed a shocking social crime. I came off night duty that morning and, because it was a very beautiful sunny day, I went back to the Forest for a walk before returning to the house to sleep. I got back at about half past four in the afternoon and found the two old ladies sitting in their drawing room having some tea. As it was Christmas, I went in to see them before sleeping. But the moment I sat down in one of their comfortably upholstered, wing-seat armchairs I must have instantly fallen asleep. I came to two hours later. It was by then quite dark, and I found that the two good ladies had with amazing kindness just sat there still, in silence, for all that time, so as not to disturb me!

That autumn of 1945, the Labour Party held its annual conference in the Winter Gardens in Bournemouth after its landslide victory in the general election in the early summer. Although I had not voted Labour, I went along to see the proceedings and how their leading personalities - Alec and Ernest Bevin, Patrick Gordon Walker, Sir Stafford Cripps and the others - performed. I was hopeful that the political turn-around, particularly the introduction of social and health provisions recommended in the Beveridge Report, and the nationalisation of the railways and the coal-mines - both those institutions being in a state of financial collapse under private enterprise - would help to get the country going again and lead to a fairer, less class-conscious society. (Most facets of that situation seem to have been forgotten by now, in the nineteen-nineties, after years of propaganda for privatisation.) The atmosphere of the 1945 Labour Conference did encourage such hopes. But only a few years later I was sadly disappointed by the rifts and disputes that opened up in the Labour Party, so that it seemed to me better to continue to support the Liberal Party even at the bottom of its fortunes in the nineteen-fifties.

The following year, 1946, was distinguished by a remarkable succession of thunderstorms and ever-recurring thundery situations, which with frequent southerly winds were especially impressive near the south coast and repeatedly brought dramatic threatening cloudscapes.

When the new London Airport at Heathrow opened in August 1946, and became the main UK air terminal, Hurn was downgraded and I was transferred to Prestwick, resuming my connexion with Trans-Atlantic flying. And very soon I and my little car were back in Scotland. I at once took the opportunity to make a first visit to the hills on the Isle of Arran. But it was not to be for long. On 4th September I received a letter from Meteorological Office headquarters, telling me that the Office was looking for a volunteer to be detached for service with United Whalers Ltd. The firm was about to take delivery of a new whale-factory ship carrying two "Walrus" amphibian aircraft for whale spotting in the Antarctic. The whales had been overfished before the war and were expected to be still scarce. The aeroplanes were to spot whales from the air, but John Grierson (namesake of a film director of that time), who was to be in command of the flight and its half dozen air crew, all ex-Fleet Air Arm, was insisting on having meteorological advice on board. Without that, he would not undertake the operation.

I later learnt that fifteen members of the Meteorological Office staff who were thought likely to be interested and had some suitable experience had been sent the same letter. I had long been very keen to see Greenland on which I had made some study and had written a report during my training period at Croydon. But Antarctica offered much of the same fascination, and I was immediately ready to settle for visiting the Antarctic first.

There was one colleague on the forecast bench at Prestwick who also received the same letter, but he was married and I had not even met my wife to be at that date. So I wrote, and sent, an enthusiastic application at once. He did not write till later. I was convinced that my experience in Ireland during the war in making forecasts for flights over an ocean, with no weather observation reports available other than what the aircraft themselves experienced, would be highly relevant to the problem of flights over Antarctic waters. Moreover, my acquaintance with the Norwegian language should be a help in the whaling community where most people were Norwegian. So I got the job.

Interlude

A few days later, I was once more heading south in my little, always heavily laden, car. Once again, with the kind support of my friends the Conollys in Surrey, I based myself with them and travelled up and down to London, beginning to find out what stage the planning of the exploit had reached, what I would have and what I would need to secure in the way of equipment, instruments, and reading matter, and making a start on buying suitable clothing, though more of that would come later from the ship's "slop chest". In many respects it was a whole new world that lay ahead of me.

Some planning had been done before I was appointed to the project - quite a lot in some directions but, as I was gradually to discover, nothing at all on other aspects. And there was a very limited number of days to go before the departure. I reached London on 13th September and fourteen days later I joined the ship on Belfast Lough.

CHAPTER 13. GOING WHALING

Whatever one's conscientious scruples about taking the lives of animals, there are some times and circumstances in which it is more easily defended than others. The nearly world-wide shortage of food and widespread under-nourishment of the human populations in many countries in the first year after the Second World War made that such a time. And, as is commonly overlooked or forgotten by campaigners in warmer countries, the precariousness and privations of life in the lands of the Arctic and the Arctic fringe have always provided justification, at least locally and regionally, for sealing and whaling by the small and sparse populations of the coasts of Norway, the Faeroe Islands, Iceland and the Eskimo lands. Such activities must surely have been bound to start in all such places around the world, both north and south, and including (as is well-known) the colder shores of the north-eastern United States, Canada, Russia and Japan.

I had been interested in polar exploration from my quite early years. Nansen's account of his *First Crossing of Greenland* in 1888 had long been one of my favourite books. And my preparations for the expedition now to go ahead to the southern whaling grounds with the Whale-Factory Ship *Balaena* included getting hold of copies of the tales of many epic journeys in high latitudes, their difficulties, and the weather experienced, by Amundsen, Scott and Shackleton in the Antarctic, as well as the useful analyses of Antarctic weather and climate by Wilhelm Meinardus and Sir George Simpson. The Meteorological Office was very generous and helpful in finding for me discarded copies from offices that had closed of some of these works. Other works I bought for myself. When I reached Met. Office headquarters in central London on 16th September 1946, I found that preparations had been going on for some time already before I was invited to join the expedition. It was clear that the directorate was keen on this project and had seized on it as an unexpected opportunity to acquire meteorological observations, and gain some insight into the workings of the atmosphere, over regions which had been little visited and where very little was really known. Dr. A.C. Best, who was head of the instruments division at that time, had been put in charge of all the planning and preparations. And a - to me quite formidable - collection of instruments, including many which I had never seen and none that I had ever set up for use - was being gathered together. With only ten working days to go before I was due to join the ship, I felt obliged to give first attention to my responsibilities with the instruments where my experience was least and where I was most diffident about my ability. This prevented me giving the attention desirable to the

Going Whaling

forecasting problem, and it was not clear whether the planners had given much (or, possibly, whether they had given any) thought to this. The question of how any weather forecast advice to the air party could be organised seemed not to have been considered, although that was the chief reason for the invitation to the Meteorological Office to send a man on the expedition. The ship was to carry two Walrus amphibian aircraft, which would be launched into the air by catapult from the after-deck, and which were to be recovered from the sea surface afterwards by derrick. It seemed to be assumed that the whaling company would provide whatever back-up staff and equipment would be needed. I did not discover until much later, when we were at sea off Africa, that nothing at all had been done to provide for this, and that absolutely no understanding had been reached with United Whalers Ltd. about what the forecaster's job would require.

During the week of 16th to 24th September I visited the Met. Office Marine Division in Harrow and discussed with the senior directorate in Kingsway, and with the Naval Weather Service, what might be produced for me in the way of working maps. As a result, I planned to meet Commander Graham Britton, R.N., at Simonstown in South Africa when our ship called at Cape Town. There was also an obvious need while in London to visit the famous "Gardiners Stores (Everything for the Sea)" in Commercial Road East in quest of clothing items. Over the following weekend I drove to Bournemouth, Christchurch and Poole to say goodbyes to friends there, including William Aldridge, who had been stationed in Lisbon during the war and, as a sailing man, also had had many contacts with the Navy. I slept that Saturday night in the Met. Office at Hurn Airport and went on after that to spend a night with my parents on their holiday at Burford in the Cotswolds. On the 24th I paid final visits to the Marine Division people in Harrow, and to the Naval Weather Branch, and then completed my packing.

On the 25th I was booked to go by train to Stranraer, on my way to Belfast, at 4.40 p.m. But first I had an appointment at 9.30 a.m. at an Air Ministry building in Holborn for another medical, in other words, for injections, the usual routine before being posted on overseas service. I arrived in good time, but the big waiting room was already full of people, all with appointments at 9.30 a.m. - a characteristically time-wasting procedure which ignores the personal problems and responsibilities of military recruits. When after an unexplained delay, the youngish uniformed RAF doctor arrived, he started seeing people one by one, calling the name of the one expected to come forward. By 11.30 a.m., there was no indication that my turn was anywhere near and the room was nearly as full as ever. Moreover I had to have time to collect a stock of blank weather maps from the Naval Weather Branch. So I stood up, and pointed out that, if I did not have time

Going Whaling

to do this and catch my train, the ship which was due out next day would have to sail without me and there would be no point in my having the injections. So, it was thereupon decided that my turn had come. The doctor proceeded to ask me if I had had a yellow fever injection lately and other injections for other tropical diseases, which I pointed out were not rampant in the Antarctic and that I might sensibly miss those jabs. He was quite shocked, but to save time - and, doubtless in order to be rid of an awkward customer - he decided he could let me go.

In this way, with the Air Ministry machine saved from choking itself, I managed to get my maps and my luggage and went aboard the sleeper for Stranraer, where I arrived at 4 a.m., sailed for Larne at 7.05 a.m., arriving there at 9.40 a.m., and reached Belfast by 10.20 a.m. After signing on with United Whalers Ltd. at lunch in the Grand Central Hotel, I joined the whaling factory ship, *Balaena*, on Belfast Lough. The deck was still covered with a mass of ships' parts, heavy chains, and other sea-going material. With all that and the continual clatter of metal objects being assembled, the impression was that the ship could not yet be more than eight or nine-tenths built. I stayed up until 2.30 a.m. next morning to see all my meteorological gear in its crates and boxes, and my luggage, safely on board. Two tugs turned us round at 10.30 a.m. and then left. At this stage, there were sea trials of the ship and of the catapult launching the aircraft, and so on. Finally, we sailed out of Belfast Lough in the late evening on the 28th. And on Sunday, 29th September, we sailed peacefully forth in brilliantly clear weather, under a cloudless sky, through the Hebridean Sea. It was an odd start to a voyage to the Far South, but we had to pick up the crew in Norway. In our northward passage through the Hebrides the opportunity was taken for further trials of the ship and aircraft and the launch of recovery gear. We sailed on north through the Minches and passed Cape Wrath at 11 p.m. Our destination was Tønsberg, Norway's oldest town, in the south-east of the country, not far from the mouth of Oslofjord. We lay for five days near the outer end of Tønsberg fjord, a wonderfully sheltered natural harbour. We looked across to the long leafy island of Nøtterøy, having come in through a narrow channel called Husøyaflaket, not far from the island of Jersøy. The area now has many pleasant homes on its shady banks. Our sojourn here was just long enough for the youngest of our dashing ex-Fleet Air Arm pilots, Geoff Collyer, to get engaged to be married to a sweet girl he met on Nøtterøy. It also enabled me to surprise my old friends, the Müllers in Oslo, whom I had stayed with several times before the war, by telephoning them. The result was a great re-union party in Oslo, at which not only did the two branches of their family that I had known turn up but our mutual friend from England, John Kellock, who happened to be on business in Oslo, came along too. We have kept in touch ever since. Also, in the Tønsberg-Jersøy area is the burial place of a famous Viking ship, the Oseberg ship, which is now specially housed in Bygdøy museum near Oslo.

There was immediately much business to do in Tønsberg, meeting the officials of the whaling company and arranging provisions and stores as well as taking on the three-hundred-and-ninety-strong crew of our whale factory ship. There is an area hereabouts as far along the coast as Sandefjord and Larvik, which has been since at least the nineteenth century one of the world's greatest centres of the whaling industry, possibly the greatest of all, and the most experienced whalers in the world mostly have their homes here. Such a mystique had been built up over the years that it was widely believed around the world that no-one could run a successful Antarctic whaling expedition unless it was led by a captain drawn from one of the well-known twenty-five or so families in this area who had the special skills and experience on which the tradition rested. And, in fact, all the whaling expeditions which sailed south in the 1946-47 season from Norway, from Britain, from the Netherlands, from Russia, and from Japan, were so led. Much was to come to light during our voyage about how jealously the tradition was guarded.

When we sailed out of Tønsberg on Sunday, 6th October, we had our full crew of 390 men on the factory ship. Of that number about 350 were Norwegian. The exceptions were chiefly the air party, the International Whaling Commission's inspector, Commander Buckle, the fifth mate, Carter, the ship's surgeon a lean Highlander called Hutchinson, the biologists Dr. Case and the young Aberdonian Dr. Begg, myself and just a few deckhands. We also had with us the Chairman of United Whalers Ltd, Mr. Trouton and his wife and twelve-year-old daughter. We called once more in the United Kingdom and lay in Southampton docks from the 9th to the 15th October. This was the point at which the builders finally left the ship. The decks were beginning to look a little clearer. I had unpacked and checked my equipment and had already been ordered by the captain to remove it - chiefly the meteorological instruments - from sight. Some of the instruments were quite bulky. They were now banned from the public rooms and I had not been given permission to set them up anywhere else. There was no office for the meteorology and I had only the half cabin aft where I was to live and sleep. Obviously, some delicate negotiating lay ahead, and I would have to get to know the captain and other officers of the ship and functionaries of the company on board. Much of the meteorological equipment was still in its packing cases and on other people's premises. Clearly, none of it could stay on the flensing deck. I was to discover that the various difficulties were symptomatic of the general unwillingness to co-operate and the lack of understanding on the part of either the ship's direction or the company, in any way to deflect time, effort or facilities from the closest attention to the whaling for any purpose whatever.

This stay in dock at Southampton made it possible for me to pay one further brief visit to Met. Office headquarters and a final visit to my parents. It was as well that I did this because on return to the ship I found a telegram from an aunt telling me that my father had died suddenly on that last Saturday evening while talking enthusiastically to another of his sisters about the expedition that I was embarked upon. He had, I suppose, too suddenly and too belatedly discovered that I was doing something that he could regard as worth while. It was a sad and solemn moment. These two aunts urged me very strongly to go ahead with the expedition and assured me that, however sadly, my mother was also convinced that that was the best course and they would care for her welfare.

We sailed early on Tuesday, the 15th, the difficult problems for me on board still unresolved. Luckily, two key characters, Per Virik, the First Officer, and Leif Larsen, the Ship's Secretary, did all they could to smooth things. Meanwhile, John Grierson, the Flight Commodore, was experiencing just the same kinds of obstruction and his log was full of similar complaints to mine. In the end, the situation was far more serious for Grierson and the air party than for me, because nothing could destroy the value of whatever meteorological observations and experience I managed to acquire, whereas the airmen could be prevented from flying or doing anything useful whatever. And for most of the time they were! This frustration of the air party members, most unfortunately but hardly surprisingly, gave rise to nationalistic reactions. What was in reality purely commercial pressures tended to be interpreted by the air crew as anti-British machinations by the Norwegian personnel. This was, I believe, completely groundless and unwarranted: the case only illustrates how easily unhelpful tribal suspicions and feelings of hostility are aroused in many people.

It also came out in various people's conversations that Captain Pedersen, and even at one point that the Chairman Mr. Trouton, had forgotten - if he had ever been told - that the meteorologist was with the expedition because the flight commander did not believe that the safety of the flying could be secured without such provision. The flight commander had stipulated that without him there could be no flying. The contention of one of the directors of United Whalers, Krogh-Hansen, who was on board, that my presence had been granted to the Meteorological Office as a privilege, so that the company had no further obligations towards me, was a bizarre misunderstanding of the situation. There was much reason to be grateful for the diplomacy of Larsen and Virik.

On the long voyage from Norway to Antarctic waters we were accompanied, at some distance, by our fleet of ten catcher vessels, like large trawlers, each with a harpoon gun mounted in the bows. One of these boats went all the way without a compass, relying instead on the radiotelephone to keep contact with the factory ship. The fitting for the compass in the wheelhouse was used instead as an ash tray! And, on the voyage south on the factory ship, I continued putting up the meteorological instruments here or there, wherever I could get permission, rather than where was most suitable for their working. I managed to avoid any further instances of being ordered to take an instrument down. Setting up the instruments was a daunting task for me. In some ways I felt like a small boy coming into a room with a great pile of Christmas present parcels, nearly every one a secret, around a Christmas tree. There was an agonizing side of this to me. Because of my father's manifest disappointment in me as a worker with carpenter's or any other kind of tools, I have become convinced that I really am extraordinarily lacking in practical skills. So I have always been prone to panic before any jobs that demand the confidence needed to produce an accurate finish. I was now confronted with a situation in which I could not know what sort of instrument would be found in most of the crates and packages until I got it out. Many of the instruments I found I had never seen or handled before, and I had never worked in the instruments division of our organisation. In the event, I was only defeated by one instrument - the raingauge. There were several difficulties with its use. Most of the voyage was in latitudes where all the precipitation was snow, not rain. That perhaps did not matter greatly, because the snow in the gauge could be melted (although there would be some loss by evaporation, in the process). The more serious difficulty was that the great factory ship constituted such an obstacle to the free flow of the wind over the ocean surface that one could actually watch the snow moving upward as the wind itself was forced upward in passing over the ship. Any measured "snowfall" caught in the gauge, which was actually exposed as far clear of the ship as possible by hauling the gauge up on a pulley to near the masthead, could not be properly representative of what was falling in the area. I never solved this, but in any case I had to plead that my time was more than fully occupied by the weather mapping and forecasting requirement.

It became obvious, before we had gone very far, that another difficulty was that the ship's radio officer, Fugleseth, could not cope with the meteorological messages that would have to be received in addition to the ship's own traffic and the personal messages for the people on board. There would have to be additional facilities, if enough information about the general weather pattern were to reach me for any sound advice to be provided for air safety to the flight commander. The only workable scheme that I believed it possible to devise was to take the weather broadcasts transmitted from the inhabited countries and islands all around the low and middle latitudes of the southern hemisphere, and then draw a great weather map - once each day - which would show the shape of the major features of the wind pattern controlling the advance of individual weather systems towards us. If one could in that way spot the shape and development of the slow-moving regions of high barometric pressure - the subtropical anticyclones in what used to be called the Horse Latitudes - it should be possible to deduce the prevailing directions of the wind flow in those latitudes from day to day which carry forward, and steer, the fast-moving storm systems and weather fronts in the "Roaring Forties" and "Fifties". This procedure should tell one, I believed, the direction from which the next disturbances to our own weather in these, and higher, southern latitudes should be expected to come. And this would be something we could check with our own observations, with what we could see in the sky and check with our aircraft, and the continually changing network of observers that our whale-catcher vessels could provide.

For this scheme to be put into operation a much better level of co-operation would be needed. But experience already showed that, not only Virik and Larsen, but also the harpoon gunner captains and crews of the smaller vessels, as well as the more junior officers of the factory, were keen enough to co-operate.

In the meantime, we called in the Canary Islands, at Las Palmas, for further provisioning and purchase of wines and port, on the 20th and 21st. The late October sky was as leaden as it so often is in England and central Europe at that time of year. But at dusk on the 21st, the cloud sheet began to clear from the north, and in doing so at last revealed the great mountain of Tenerife, 12,200 feet high (3,718m). Although visibility was really very good, there was a light haziness about the dusk and there crept into view that whole huge mountain - like the unveiling of a mystery - seven or eight thousand feet of it, riding above, and as if based upon, the retreating cloud layer: as it were, a false horizon seeming to support a huge mass. It was an unforgettable moment.

Las Palmas has its share of grand architecture, like so many south European towns. But at that date it looked somewhat "down at heel", with young Franco soldiery standing about the street corners in the port area sheepishly, while little mobs of 12 to 14-year-old boys tracked us and accosted us as we walked into the town in the evening. One of these succeeded in selling a bottle of "brandy" to one of our number who later found it to be just coloured water. Sunday evening was a fiesta, and young choir boys in surplices, who were scarcely into their 'teens, carrying the image of the Virgin through the streets stick in my memory. In more detail, what I remember is those boys smoking cigarettes and continually scratching itches on their legs beneath their bright red cassocks.

We sailed at dawn on the 22nd. I was acutely uncomfortable for the four days of heat as we made our way south through the Doldrums belt, near the equator. But the giant cloudscapes were interesting: long lines of great thunderclouds, one line with the slight curvature characteristic of a trailing cold front from the North Atlantic, and the other line nearly converging with it, having the characteristic curvature of a slow-moving cold front that had come from the southern hemisphere. (These characteristic curvatures are the result of stronger winds associated with cyclonic depression systems in higher latitudes that have passed east along the front, while that part of the frontal system that is in more sluggish winds trails behind the disturbance that has gone on farther east.)

At this point, I decided there was no more time to lose before bringing matters to a head over the facilities on board for doing the meteorological job. So, as we neared the equator, I drafted a telegram to Meteorological Office headquarters in London, saying "Suggest re-call while *Balaena* is in Cape Town, as no facilities provided for my task". The telegram never went, but it did the trick! Nevertheless, to make matters worse, the company chairman, Trouton, reportedly suffering from insomnia, now complained to me that, since I had been training the flying personnel to act as supplementary meteorological observers, they disturbed him in his cabin by chatting with the ship's officers on night watch on the bridge. To me, this came as unexpected evidence of tension on board rising all round. The same day, Captain Pedersen was enraged by the flight commander, Grierson, setting up a telephone that I had lent him to communicate with the aircraft for flight plotting practice. This led to me being ordered to take my weather charts and other gear out of the smoke room which had come to be considered as an interim Met. Office, and therefore allowed on sufferance up to this point.

Things had come to a head in more ways than I suspected and the underlying reasons for tension were beginning to come more clearly to light. At the bottom of much of it, was the Captain's resentment at all the scientific and technical aids to the search for whales. He did not need the aircraft, or the air party, or the meteorology. He, like all his coterie among the traditional whaling families, knew how to find the whales and he knew what the weather would be. His position was, of course, at cross purposes with the company's. However, things had been set in motion by my draft telegram, and with Grierson's support I got a decision from Trouton that a

Going Whaling

big empty room above the flensing deck amidships should be furnished as a Met. Office cum flying control room. And an extra radio officer should be engaged solely for the weather messages. All this was to be done, and a vital visit was to be paid by me to the Naval Weather Service station at Simonstown and its chief, Commander Graham Britton, during the five days from 6th November that we were to be berthed in Cape Town. The meteorological instruments could stay in the positions where I had by now managed to set them up, because vibration from the ship's winches would be prohibitive in the newly designated Met. Office. This was actually a good-sized, comfortable room though it suffered from a drawback in being awkwardly far from the aircraft hanger aft. There was a further difficulty arising from its position (which might have been better planned beforehand) that access was only by a steep ladder up from the middle of the whaling deck, and the hand-rails on either side became unavoidably contaminated by whale blood and fat. This meant that the hazards of the deck spread up the hand-rails, and they had to be treated with the greatest caution. In the course of the whaling season, two deck-workers died of gangrene, contracted on deck, and had to be buried at sea. Nevertheless, one had to recognize that progress was being made and be thankful for it!

When we came to Cape Town, the result was that we recruited a young man, Robert Currie, from Watford, who happened to be there and unemployed at the time. For him the job carried the bonus of a free trip home. While there we stocked up with good South African sherry and liqueurs. Van der Hum was the one I liked best. It was also presumably the favourite of a deck-hand who paid an unauthorized visit to my cabin one evening over Christmas and stole a Van der Hum bottle from my locker. When empty, I had used it to mix ready for use on the weather maps the stock of government supply red ink powder. Next morning, our one and only Cockney deck hand was in the queue outside the doctor's cabin. He was very worried after wetting his bunk. He seemed to have passed blood. The exact solution to his problem only dawned on us in conversation with the doctor afterwards in the ward room at coffee time! We never discovered if the man still liked Van der Hum.

I went for long walks through the city in the sunny evenings, passing through both white and coloured districts. I saw no disturbances and met no hostility. That was in November 1946, before the days of the Apartheid governments. I also went with a party from the ship up Table Mountain by cable car, dangling at times high above the ground. Deeper thoughts occurred on the way up, and I decided to come down the mountain on foot, as there seemed little likelihood that the German-built cable-car system had been properly inspected during the years of the war.

Going Whaling

We sailed for the South on the 11th, and headed south-east from the Cape, to face over five months at sea, an experience quite rare among mariners and not exceeded by many in the Royal Navy. With my newly appointed radio operator, Bob Currie, I occupied the spacious new office on the 21st. As our ship headed into the usual zone of the Brave West Winds, between latitudes 40° and 50° S, we experienced the sort of weather that every meteorologist must be prepared for - "the exception that proves the rule". Our wind settled into the east, seldom very strong and only briefly approached gale force. Later it became northerly and there was just a little rain. Whaling began with taking sperm whales on the 17th. Sperm whales provided the raw material for one or two skilled carvers on board to make "penguins" out of their teeth, a potentially profitable tourist trade. But we did not continue with the sperm-whale fishing for many weeks.

Continuing difficulties with the captain are registered in my diary of many conversations. It became obvious beyond any doubt, that he regarded the whole air operation and the weather forecasting enterprise as a threat to his own position and to the industrial traditions of the Tønsberg region. However, competition within the industry made its presence felt on the whaling grounds, even so far from home. Already on 23rd November, during the sperm whale fishing, Captain Pedersen was complaining to London by radio that the other Norwegian whalers on the whale factory ships Norhval and Thorshammer, which we knew to be operating not far from us - and on one day, the 25th, briefly within sight - were not broadcasting any weather reports and, if that continued, he would soon stop sending reports too, so as not to give away our position.

On 1st December we received a very co-operative telegram from the Naval Weather Service at Simonstown, announcing that they were to broadcast their weather forecasts for the whaling grounds and messages, giving collected weather reports from the South America, Falkland Islands and Falklands' Antarctic Dependencies sector, for our convenience on account of difficulties we were having with radio reception. At this date whaling was going very well and both Captain Pedersen and Trouton were in affable mood.

From about this point on, conversation in the mess rooms increasingly dwelt on how the catch was going and the likely bonuses that would be earned by the end of the season. Others among our companions at sea proposed that fines should be levied on those who "talked shop" in this way too often. In fact, over the season, there was much very good conversation, at least in the senior mess room, ranging widely over geographical, historical, and scientific exploration and international affairs and much else besides, including naturally individual exploits and achievements, marital and family affairs. At Christmas and New Year we had festive meals in both the British and Norwegian traditions and ended with sing-songs around the tables, led by individual performers, including ballads from Norway, Sweden and Denmark, England, Scotland and Ireland, Germany, America and Cuba. The repertoire included many popular traditional seafaring songs from the days of sail. We enjoyed a prolonged holiday from wartime and post-war food rationing and did not need to eat much whale meat after the chances the ship had had to stock up in Britain, Ireland, Norway, the Canary Islands and South Africa.

The "inner man" was well catered for and I, for one, much enjoyed the simple things: the always freshly made pale, Norwegian liver sausage, the Norwegian cranberries served with most kinds of meat (notably mutton) - always cheering for their bright red colour and sharp taste, the freshly baked Wienerbrød (literally Vienna bread, though actually a staple delicacy of the Danish and Norwegian table), and of course the splendid coffee available at all times in the ship's galley: it was especially welcome as a restorative before returning to the work-room at midnight to complete my southern hemisphere weather map and forecasts for the day. One Norwegian Christmas speciality which I found very much to my taste, but which did not appeal to many of the British on board, was *lutefisk*. It is made from portions of white fish, long soaked in lye (alkali) and partially dissolved in the lye until they become glutinous. They are then served with melted butter. This came as a starter before the main course of the Christmas dinner consisting of roast pork and prunes (also very appetising). There was, however, one aspect of the whaling season which everyone - even the toughest constitutions among us - found really hard to bear: the stench in the factory. Even the toughest characters among those who were new to the whaling grounds eyed each other anxiously in the first days of the season, wondering if they would manage to see the season through. The test was severest, as could be seen from their faces and heard in their comments, when one met them below decks in the area where the distillation boilers were.

In mid season we had a visit from an oil tanker to refuel the factory ship and, after thoroughly cleaning out the boilers, to take off our first whale oil production. The tanker vessel also had an important mercy mission to fulfil, by taking off one of the flight engineers who had never mastered his sea-sickness. This man was started on his way home to Europe, as a passenger on the tanker. When this tanker arrived, she was a beautiful sight with the white painted upper decks gleaming in the clean Antarctic air and the southern summer sunshine on the February day when she arrived from the Caribbean. The wind was a brisk southerly, bringing fresh, cold

Antarctic air. We avoided the swell by keeping close to the pack-ice edge so that the air had not passed over any open sea before reaching us. The tanker was lashed alongside our factory ship with several recently shot whales inflated by compressed air floating between the two big ships, to serve as fenders. (Of course, decay was not rapid in that cold sea, so there was no smell of decay in the open air.) But by the second or third day of the tanker's visit her beautiful white paint had been transformed to a blotchy daub of colours ranging from indigo to copper by the powerful tarnishing effect of the effluent from our boilers. Visits were paid from ship to ship, while they were lashed together, by parties of up to three men carried in a bosun's basket slung from a derrick or by individuals putting one foot into a stirrup and travelling with a pulley along a cable between the ships. This was a thought-provoking experience as one contemplated the two thousand fathoms or more of ocean beneath one.

There were other facts about the arrangements for this tanker's visit that threw a memorable light on how the whaling industry had been carried on from time immemorial down to the period we are here concerned with. For a couple of weeks before her arrival our skipper had mentioned on several occasions that there was a fjord on the coast of Antarctica which he planned to use for shelter while the ship-to-ship transfers were going on. But when the time came he could not find this feature. The explanation for the difficulty was not hard to guess. The valuable knowledge of the fjord and its shelter had been treated as a trade secret, never published and never allowed to reach the map-makers! As late as 1946-47 the coast of Antarctica was entered on maps only as a broken line, with no hint of any detail. Features such as the fjord that Captain Pedersen was looking for may have been discovered many times but never reported to the map-makers. The information had only been passed by the whalers from one to another among their cronies back in Norway in the summertime and never, never publicised.

The every day scenes of carnage on the deck of the factory ship were horrifying. No one need doubt the lethal risks to human life lying around. The catcher vessels delivered their whales, sometimes two at a time, as inflated corpses made more buoyant by the injection of compressed air. The corpses were then moored at the stern of the factory ship until there was room on deck to haul them, one at a time, up the broad ramp out of the water. As soon as the whale was on deck and held in place by stout metal hawsers, one of the whale men would be up standing on its back with his flensing knife (shaped rather like a hockey stick), starting to peel off the blubber, four or five inches (ten or more centimetres) thick. Our biggest whale of the season, a blue, was 96 feet (29m) long and 13 feet (4m) thick, at its thickest point. The blubber was cut down into strips and soon consigned to the boilers by way of open manholes in the deck. Then the bones were sawn through by mechanical saws and, in sections, ground down in the factory for bone meal, to be marketed as fertilizer. The guts were quickly winched across the deck to the side of the ship and into the sea by a single hawser around the back of each lot. Needless to add that the deck became soaked in blood and for that reason alone, apart from the risks of the open manholes to the boilers (and the slipperiness wherever there was a bit of metal surface), it was a place of lethal dangers. One of the fatal accidents was caused when a metal hawser parted under strain, and sprang back catching a man and cutting him in half. The other death resulted from an injury with a flensing knife. In the burials at sea that followed these tragedies, the Norwegian Lutheran service was read by one of the ship's officers, the language being very close to that of the English Book of Common Prayer. The ceremonies were attended by nearly everyone on board. On the homeward voyage, in the first days after whaling ceased, the blood-soaked boards of the deck were lifted and winched overboard into the ocean.

Blue-whale fishing had begun, in compliance with the International Whaling Commission's regulations, on 8th December in latitude 56 to 58° S about 75° E. The struggles with the ship's authorities continued in spite of the good start to the season during the sperm whale fishing. Permission was never granted for any meteorological instruments to be set up in the wheelhouse, despite the convenience of such a position to all concerned with using the instruments, and although all the officers of the watch agreed that the wheelhouse was the one obvious place that was in shelter and away from vibrations. At last, on 15th December, the chairman, Trouton, got me and Captain Pedersen together and tried to get him to warm up and co-operate and "tell what he knows" of the weather on the whaling grounds. But Captain Pedersen was plainly shy of doing so. My pleas to be shown any messages received either from the catchers or from ships approaching our fleet from Cape Town, which could be of great value to the weather analysis and forecasting, were turned down on the ground that the ship's management could not spare the time, despite the support of most of the ship's officers. By the end of December, relations between the air party and the captain were becoming worse than ever. There had been no flying since 4th December. On 1st January we had the most worrying incident of the whole season. A flight was launched despite fog patches and bad visibility due to an advancing line of snow, all as forecast. The pilot, Geoff Collyer, doubtless anxious to demonstrate the value of the air support and expecting some co-operation from the factory ship in his effort to avoid the snow-belt, which was advancing at about 10 miles an hour, climbed to five thousand feet (1,500m) for a better view. But the factory ship did not change course until too late. The aircraft lost sight of the ship and made an emergency landing, alighting on the water in light wind about twenty miles away. Two whale-catchers were sent to look for him and guide him taxiing back to the ship. The captain was furious at the loss of two hours whaling time and told Grierson that his aeroplanes were "nothing but a b---- nuisance". Other developments soon followed. On the 4th the captain was confined to bed with a stomach ulcer. So investigation of the flight emergency was delayed. On the 5th Grierson and the air party announced that they had no intention of carrying on. They proposed to return home with the oil tanker, *Norvinn*, expected later in the month. On the 5th Grierson conveyed this decision to Trouton.

Later the same week, the Captain gave leave for my forecasts to be relayed to the ten whale catcher vessels, since the gunners were keen to have them. Trouton urged me to emphasize the geographical distribution of the weather in my forecasts, to issue them about midnight to 2 a.m. ship's time and to add an outlook for the next 24 hours ahead. All concerned were asking for a clear expression of the degree of confidence attaching to each forecast. Trouton then showed me his analysis of the weather and working problems during a previous year's fishing season. Apart from poor visibility, the main weather factor affecting catches was really strong wind and seas disturbed enough to make it difficult to secure whales once shot. Fishing would continue in anything less than a strong gale with prohibitive seas. But if whales were scarce, visibilities less than 5 to 8 miles (8 to 12 km), whether from fog or snow, restrict catches.

Around 12th January I found that my forecasts seemed to be deteriorating owing to lack of sleep after working till 3 a.m. ship's time. These hours of work also upset my meals routine and meant that it was easier to have two dinners in the day but miss breakfast. My routine needed changing.

On the 13th January Trouton left, with the oil tanker, *Norvinn*, to return to the company's office in London. The captain was still in bed with his stomach ulcer, and was given a blood transfusion. It was noticeable that during the two weeks of his illness, the co-operation that both I and the air party received was the best since the voyage began. On the 15th Chief Mate Virik put forward the ship's clocks by five minutes, so that the beacon for the catcher vessels homing towards the factory, which broadcast each hour at the hour and half hour, would not clash with internationally broadcast weather messages which we needed to receive. On the 19th Director Krogh-Hansen told the air party that their flights in the last two days had been very useful. "Such flights should be repeated almost every day". And, on the 25th January, the star gunner, Finn Ellefsen stated that "the weather forecasts have been surprisingly successful and a very useful guide". Later, he visited the factory ship to tell me this in person. He wished to be able to telephone the factory ship at any time, to secure weather forecast advice. During February he further informed me that this was the general opinion of all the harpoon gunners in our fleet.

In mid February, I managed to take a two-day break on Ellefsen's catcher and saw how they operated, the firing of the harpoon gun, the paying out of the line to the harpoon which held the whale, followed by the inflation of the carcass with compressed air, and flagging of the floating carcass, while we went on within the range of sight of our flagged whale to chase other whales. The weather during my trip was excellent, with unlimited visibility and bright sunshine throughout, on the water, the pack-ice and bergs. The temperature was -9° C through the day, and I was amazed by the enthusiasm and endurance of a young Norwegian deck-hand on the catcher, barely 20 years old, who spotted a chase going on while he was crossing the deck in his jersey and jeans to put his mug away in the crew's quarters below decks for'ard. Without extra clothing, and with bare hands, he stayed on deck for half an hour to watch the developing chase and the general scene, which was one of the most stimulating we had had. The brilliant sunlight and supremely clear air made it very beautiful and exhilarating, but the suffering of the whale with its blood discolouring the ocean surface was not a pretty sight, and the whale's efforts to escape by plunging deep were hard to watch, though luckily not very prolonged in the cases I saw.

My trip was made possible by the kind co-operation of Leslie Holmes of the air party who took over the meteorological observations for 48 hours while I was away. All the air party were by then "champing at the bit" for a chance to prove what they could do.

I thanked the captain a few days later for having readily given me permission to go with the whale-boat and told him how much I had enjoyed the experience. At this, he smiled at last and said affably: "Of course, you could not come down here without having a trip on a catcher".

On 20th February, however, we learnt from Trouton in England that the company had decided not to take aircraft for whale spotting in future years. The hangar would be converted into cabins for a hundred extra workers. This was, of course, a bitter blow to the air party, whose position had been fundamentally undermined by the unexpected abundance of whales.

Going Whaling

Some geographical discoveries of previously unknown details of the Antarctic coast had been made, but the whaling company was still not keen that such details should be published. I myself had sought the captain's permission on 22nd February, when we were near 65° S 111° E, to signal the Air Ministry in London notifying land which I had sighted that day from the ship and taken bearings on. It was a part of the coast of Antarctica, which had first been sighted long ago and named Budd Land on the charts in the nineteenth century, but it had subsequently lapsed to doubtful status on maps, after the 1939 voyage of HM Survey Ship *Discovery*, which reported it "not found". Our captain forbade me to send any such signal, adding obscurely that I "could not have seen any land", but he knew there was land there and he even knew of a fjord in about our present longitude, which might provide useful shelter from an easterly storm!

As the season went on, I continued to get increasing support from the ship's officers and particularly the radio officer and the gunners, who were the skippers of the catcher vessels. Even Captain Pedersen was at times encouraging, though always guarded, particularly regarding his superior knowledge of the weather of the region and his ability to forecast it. For that he needed no weather messages and always refused to look at a weather map, though he did believe in having occasional chats on the radio-telephone with other whale factory skippers, who were cronies of his, on the rather few days when there were any within radio-telephone range.

The pitfalls surrounding his stance were cruelly laid bare on 22nd February 1947, in latitude 63° S, when we enjoyed twelve hours of lovely weather, with light wind (force 3 to 4) and sunshine. At the end of that day Captain Pedersen tapped the barometer on his cabin wall - it was rising a little - and announced to those present that he was "quite sure we would have a whole week of fine weather now, at least six days more". Six hours later the wind was force 7 and rising, then soon force 9 from ESE. That went on all day, and the following day produced storm force 11. This was enjoyed by the disgruntled air party, who were allowed only three flights between 26th January and 8th March, as a grim joke. But the captain's attitudes were changing, however belatedly. Dr. Case, our senior biologist, told me on 9th March that the captain had remarked to him that I was "not a bad fellow in many ways" and "would learn in time about the weather down here".

The aircraft were used four times in March, each time for ice reconnaissance after stormy weather. We were now operating in latitudes near 65° S. Another new leaf was turned when Captain Pedersen, for the first time, attended my pre-flight forecast conference, which began now to be held on the bridge. Naturally, Grierson as flight commander, and the pilot chosen for the particular flight, as well as the officer of the watch, were also there. As I went through the forecast, reading out each item and discussing it, the Captain repeatedly interrupted to give his own opinion on each and every item. Notably, he made light of the expected snow and hail showers (which would matter far more to the aircraft than to the ships) and of the prospects of ice formation on the aircraft. Interestingly, because it was something not at all expected from previous ideas and experience of polar climatology, we were witnessing cumulo-nimbus clouds - the basic type of thunderclouds, sometimes congested with broader expanses of frontal cloud - advancing towards the coastal hills from the interior of Antarctica. This was the most turbulent cloud development we had seen in the whole season. By the end of the conference, the captain's interventions had so confused the discussion that we had a shorter, repeat briefing in the hangar before take-off.

We were driven to the conclusion that Captain Pedersen had very little knowledge or understanding of meteorology and regarded me with fear and suspicion as presenting a dangerous rival to his supposed expertise. His hostility to the whole air operation could be explained in the same way. It was noticeable, however, that he became more genial for a time when Trouton had left the ship. His mood underwent a similar change with the departure of Krogh-Hansen, the other representative of the United Whaling Company directorate, on the second oil tanker, *Sysla*, to take our oil production, as the end of the season approached. It was becoming clear that the captain had his pleasant and kindly side, but had felt himself under siege throughout the voyage from all the rival authorities and experts represented on board. He was undoubtedly quite an autocrat and in a position where, according to the traditions of the sea, he had a right, and indeed a duty, to command. By all precedents in his experience, a whaling captain should be the undisputed king on his own ship, particularly on the remote waters south and east of Cape Town. He may well have imagined that even the company's chairman, Trouton, would be effectively a prisoner on board down there. Pedersen, himself, was also in a lonely position with no natural allies among the challengers to his authority on board.

On the return voyage through the Roaring Forties in mid April we had winds and seas as stormy as they are generally supposed to be. The factory ship had to reduce speed drastically when derricks broke loose and crashed about the deck. About a day and a half before reaching Cape Town, we enjoyed a return to warm air and prolonged sunshine. And then, in port, from 20th to 23rd April, we rejoiced in renewed access to fruit, fresh vegetables and such things as avocados, which were a new experience for me at that date.

The voyage on home to Europe proceeded peacefully, with most of the ship's company sunning themselves on deck. My lot as the meteorologist, however, at this stage was still to spend my time making the meteorological observations and packing up the instruments. One remarkable occurrence was that when, after crossing the equator, we passed through our first northern hemisphere cold front, on 7th May 1947, around 29° N 14 to 15° W, a little south of the latitude of Madeira, the air became so much clearer and more transparent to the whole spectrum of the sun's rays, that a queue of people suffering from sunburn quickly formed outside the doctor's cabin - even though they had been continuously exposed to the equatorial sun.

We docked at Southampton in fair weather on 11th May, and I joined my mother in Cambridge on the 12th. There was still much to be seen, particularly in the Fenland, of the great floods that had accompanied the thaw in March which ended the historically severe, snowy winter that had brought many harsher experiences and lower temperatures than we had had in what, after all, was an Antarctic summer and autumn.

The captain left the ship and the company, when they reached Norway after the *Balaena* had gone aground several times in the North Sea. Maybe he had lost the taste for further experiences such as we had shared with him. He was reported to be doing the rounds on his bicycle the following year, selling insurance around south and east Norway. The air party disbanded and did not sail to the Far South again, although Geoff Collyer did marry his Notteroy bride. I felt obliged to advise the Meteorological Office, when consulted, not to send a meteorologist south again with the whaling company unless much better co-operation from the ship's management could be guaranteed.

The whaling had in fact been much better than expected, and that was the main cause of the abandonment of arrangements for whale spotting from the air. For me the expedition had been a grand experience. A great variety of valuable scientific results in the realms of meteorology, oceanography and geography came out of it, not to mention the lessons regarding human behaviour and relationships.

One meteorological lesson was about the nature of the intertropical front, or convergence zone, in the atmosphere where the wind currents and circulations over the northern and southern hemispheres meet. The unsatisfactory state of previous knowledge of this feature had been borne in on me near the beginning of my work in Ireland on the Trans-Atlantic flying, at a pre-flight briefing at Foynes in the autumn of 1939. Aircraft were flying westbound by a southern route to avoid the frequently strong westerly head-winds that were met in latitudes near 50° N on the direct route to America. They were going by Lisbon and Bathurst in west Africa to Trinidad and the Bahamas en route for New York. At the forecasting office in Bathurst one of our flight captains told me he had quizzed the forecaster about a bold line on his weather maps in very low latitudes, where there were no observations entered on the map. "What is that?", he had asked. "The intertropical front", was the reply. "How do you know it is there?", he then asked and got the reply: "Well, it has been there for the last fifty years. I don't know any reason why it should not be there tonight". It so happened that, on our southward voyage an ordinary North Atlantic cold front had overtaken us, moving south near latitude 42° N, off Cape Finisterre in north-west Spain on 17th October 1946, and its cloud system had remained more or less always in our view until we finally caught up with it again, and passed to the south of it once more, near latitude 9° N in the equatorial zone, early on the 26th. The speed of advance of the front had varied during the time that we were within range of it and its cloud system underwent a succession of developments. At first, after overtaking us, the front seemed to be going ahead more and more slowly, and its associated cloud development was becoming flatter, until it became the cloud-sheet that had caused the gloom and cut off from view the great mountain of Tenerife in the Canary Islands. About that stage, there was some more massive shower cloud development over the mountains, penetrating the flat cloud layer. Then the frontal cloud complex had accelerated again in its southward advance and moved on ahead of us. At our final meeting with the system it had become a line of thunder clouds of tremendous height. Four hours later, as we sailed on east, or east-south-east, of it our ship lay between that cloud line and another, similar one some fifteen miles south of it. And as we viewed these two lines of great thunderclouds, we could see - thanks to the very clear visibility over the ocean in those low latitudes between the cloud lines - that away to the east of us the cloud lines curled gradually away to the north and south respectively. In this way, they showed the characteristic curvature of fronts that had come from the northern and southern hemispheres.

On our return voyage in April 1947 a similar pattern was again experienced, although the fronts which we then passed were not in the very same latitudes nor in quite the same condition as when we went south. Coming north towards the equator through the South Atlantic, mostly in sunshine and under the typical fair weather clouds of the Trade Wind belt, gradually growing bigger towards the equator, we overtook several belts of light showers, this time presumably weaker cold fronts from the higher southern latitudes. We passed the last one near latitude 3° S. After that; through the Doldrums belt, the heart of the zone of convergence between the systems and air masses from south and north, this time there was nothing but cloudless sky apart from thin white veils of high cloud. So, on this occasion, not the simple arrangement of a single well marked inner tropical front never varying its position, as supposed in the nineteen-thirties, but a continually fluid situation maintained by the flow of disturbances originating from north and south towards each other.

A rather similar correction to previous ideas about the convergent flows in the water of the oceans was forced on us by what we observed from the ship. There are considered to be normally two marked convergence zones in the oceans south of the southern hemisphere continents, marking sharp boundaries between waters of different origins. Sailing to the Antarctic, one expects the sea to become colder - and the prevailing weather with it - in two distinct steps. I had always been surprised by the then generally held belief, repeated to me in 1946 by no less a person than Dr. George Deacon, founder of the National Institute of Oceanography (who later became Sir George Deacon), that those convergences never varied in position. It was true enough of what we found on our outward voyage, when we passed the Subtropical Convergence at 25° E as a sharp feature near latitude 41° S. The water surface temperature fell from about 20° north of that to around 10° C south of it, and the air temperatures we experienced after that never exceeded 10° C until our return at the end of the season. About two days later we crossed the other great convergence line, the Antarctic Convergence, the surface water temperature having sunk more or less gradually during those two days until it levelled off near 47° S at between 0° and +3° C. There was some hint of an intermediate step at which half the temperature fall took place. Our air temperatures from that point on mostly ranged between +1° and -3° C, but were occasionally lower when the wind came as a fresh southerly off the Antarctic continent. On the return voyage in April, in about the same longitude as when we were southbound, we experienced five or six more minor changes at latitudes between 50° and 36° S. It appeared that the temperature boundaries in the water surface during that season, varied in sharpness, and even in number, like the fronts in the atmosphere.

We also made some strictly geographical discoveries. Some of them were quite easily made because, from 20th February until 12th March, the coastal mountains of Antarctica between longitudes 110° and about 113° E were in full view from our ship on a number of days. They were never particularly near, always beyond the effective range of my photographic equipment, probably they were generally 25 to 30 miles (or about 40 kilometres) south of us. On 8th March our aircraft flew to a row of three, low rugged peaks that rose to perhaps 400 metres above sea level. These hills had been prominent in the view and are seen in the sketches of the coast that I made, but which were only possible with the use of the ship's large-magnification binoculars. At that date long stretches of the coast of Antarctica in the sector south of the Indian Ocean, Australia, and the Pacific Ocean, including where we were, were not printed on the Admiralty charts nor any other published map or, if shown at all, were printed as no more than a broken line. One can probably deduce from Captain Pedersen's conversation that various parts of the unknown coast had been seen from time to time, and perhaps many times, by whalers, sealers and others. By far our best view of a long panorama of the coast and coastal mountains was from a viewpoint about 65° S 112° E at 4.30 a.m. ship's time on 8th March 1947 (22h. GMT 7th) in very clear, still weather of the kind that makes the distant view seem quite near. The sky was overcast with veils of high cloud that suggested a threat of snow to come later. Peaks, some of which towards the eastern end of our view were shapely, well pointed eminences, which probably rose to heights of the order of 5,000 to 6,000 feet (1,500 to 2,000 metres). Several other massifs could be seen on bearings of 140° to well past 200° from our position, implying that we were seeing probably a 40 to 50 miles long stretch of the coastland. I was woken to see the view by the Fifth Mate, who was on watch on the bridge. He rightly imagined I would want to sketch it, but having worked particularly long hours the day before, and only turned in to sleep about 3 a.m., I decided I must leave it until I had had some more sleep. I woke and returned to the task about 6 a.m. only to see cloud fragments forming all along the seaward slopes of the high ground in the light north-easterly breeze that was developing. Only a few minutes later the whole scene was veiled in fog and low cloud and the view could no longer be recorded.

Of wider interest, however, and certainly far more exciting to me, were the indications of major geographical features of the then unknown interior of Antarctica which could be gathered from analysis of the daily weather maps. Repeated failure of the cyclonic depressions from the Southern Ocean to progress beyond certain points along the coast, or their inability to penetrate more than a little way inland, as well as the frequencies of southerly winds off the coast in different longitudes, suggested the existence of a high barrier in the interior of the broadest part of Antarctica affecting the winds much as the Alps and the Rocky Mountains do. One could presume that a barrier of similar magnitude would be needed to have the effects observed. The barrier seemed to lie between about 80° S 80° E and 70° S 97° to 100° E. This was verified only a few years later by expeditions on the ground, which reported the main crest of the continental ice-cap in about this position. The basis for this geographical prediction was stated by me in the Royal Geographical Society's Journal in March 1948:

"..... wind currents and frontal boundaries (between them), giving extensive cloud sheets, at least in the upper levels, as well as much lower-level cloud, and light snow, were regularly swept around the rear of the depressions in the Ross Sea-Wilkes Land sector (i.e. south of New Zealand and Australia, roughly between longitudes 180° and 120° E), right inland on to the edge of Antarctica's ice-cap, and later came offshore again, travelling northwards over the Wilkes Land coast. They seemed never to get farther west over the plateau than a line running north-east from about 80° S 80° E towards the coast in 100° to 102° E. The cloud-sheets and .... snow-belts were observed moving out to sea (from the interior) .... Observed prevailing wind directions near the coast both east and west of 100° E would also fit the existence of a major barrier along this line .... which would (divide in two) the territory of East Antarctica between the west coast of the Ross Sea and the Atlantic. ..... The whole trend of recurving (clockwise curving) depression tracks, both from the Ross Sea-Wilkes Land sector and in the Indian Ocean sector on the other side of this line, lends support to the belief in a topographic barrier in the interior of the continent ....say 12,000 to 15,000 feet in height."

This was not the first time that a meteorologist had been able to argue features of the geography of the interior of Antarctica mainly from meteorological evidence. The German geographer and meteorologist, Professor Wilhelm Meinardus of Göttingen (1867-1952) as early as 1909 rightly diagnosed Antarctica as the loftiest of the continents, with an estimate in 1938 of about 2,250 metres (about 7,400 feet) for the mean height of the landmass. He also estimated at that time that the main crest of the great ice-sheet reached about 4,000 metres (13,000 feet) in East Antarctica. His assessment was based on precipitation considerations and the probable shape of the profile of the ice dome, allowing for the flow that would develop.

After the *Balaena* whaling expedition I was attached to the Marine Division of the Meteorological Office for about a year to write up the results. During that time I was invited to spend three weeks in the Royal Netherlands Meteorological Institute near Utrecht, where I worked with the young Dutch meteorologist, Dr. Ben Lopez Cardozo who had also been in Antarctic waters, on the Dutch whaling factory ship "*Willem Barendsz*", during the same period that I had spent on the *Balaena*. By making a new set of daily weather maps using the observations from both whaling factories and other fleets' observations obtained after the voyages from Commander Britton's office at the Simonstown naval base we were able to produce a better analysis of the weather situations, based on many more observations from different sectors of the Southern Ocean that rings the hemisphere. The point of this was to check the errors in the weather analyses on my original weather maps by comparing with the maps now produced from the full information. The results were generally gratifying. The most awkward feature of the original working maps analyzed during the whaling season was the liability to position errors of the weather systems that affected us. The over-all average error of position of the fronts and strong wind currents was about 150 to 200 nautical miles. The practical consequences of that error margin were, however, adequately catered for by our routine practice of first sending the aircraft out directly towards the most threatening weather system at the beginning of each exercise. Having by that means determined the extent of clear weather in the least favourable direction, we could judge the length of time available for flights connected with the whaling. It was undoubtedly that that enabled us to complete the season with a clear safety record.

Other points of interest emerged from consideration of the whole map series, despite the fact that only one summer and one autumn were covered. One very plain conclusion was that the South polar anticyclone was not a constant feature present in every day's weather situation. Nor were the South polar anticyclones that did occur ever very intense. The barometric pressure levels in the central part of the anticyclones over the high ice plateau and the mountains in the interior of Antarctica were, of course, unrepresentative of conditions at sea level because of the fiction involved in correcting the atmospheric pressures indicated on the heights of the interior to sea level by adding on an allowance for thousands of feet depth of atmosphere that simply was not there. But it was clear that the pressure levels were really always low by comparison with northern hemisphere experience. Sometimes the situations over Antarctica were plainly cyclonic, just as we know the situations over the Arctic sometimes are.

The daily map series were also used to derive a map of the frequency of fronts over the higher latitudes of the Southern Ocean and over Antarctica. This showed that fronts frequently occurred, as expected, over the main storm belts of the Southern Ocean in latitudes between 40° and 65° S. But fronts were also found, and especially frequently, near the east side of the Ross Sea and over the interior of Antarctica west of the Ross Sea, including much of Wilkes Land between longitudes 100-120° and 150° E. The preferred configuration of fronts over Wilkes Land seemed to be such that the extraordinarily high frequency of fierce gales in the cold air draining off the ice-cap to the coast of Adélie Land near 140° E, first reported by Mawson's Antarctic Expedition in 1911-1914, may be associated with constriction (channelling) of the cold air flow between the fronts and the eastward slope of the ice-cap between longitudes 120 and 140° E.

CHAPTER 14. HOME FROM THE SEA; LOVE AND MARRIAGE

On coming home from the sea, my first moves - in response to general requests - were to shave off my beard and destroy the clothing that was irredeemably impregnated by the whaling smells. I myself sympathized readily with these requests after the well remembered experience from my visit to Iceland in 1938 when we picked up passengers in that whaling station at Siglufjörður, that remarkable deeply cut fjord in the north coast of the island. Like other places I visited in Iceland, the mountain- sides looked new because they were so little eroded. Geologically speaking, the mountains there doubtless are relatively new, the product of volcanic activity. And Siglufjörður is a nearly straight-sided trench, filled on many days of the year with the vapours from the whale-oil factory. And the stench was with us still all the way to Reykjavik. So I got rid of my whaling clothes before hurrying on my way to Cambridge, to be with my poor, dear, bereaved mother for two weeks. I found that, at almost seventy years old, she was bearing remarkably well the anguish of her widowhood, which had begun so suddenly just when I, her only child, had left for the Antarctic.

I brought her a souvenir of that exploit, which she cherished, and kept on the mantelpiece in her room till she died twenty-two years later. It was a beautiful model penguin, carved by one of the whalers from a sperm whale's tooth - a sort of ivory. The making of such souvenirs was a steady industry followed by one or two really skilled producers among the crew.

In mid May 1947 the Cambridgeshire fenland still presented a remarkable scene with the vast extent of the floods that followed the spring thaw that year, after the unusually heavy snowfalls of the preceding winter. By the time of my next-but-one visit, in late June 1947, the splendid summer of that year was well under way, with its cloudless skies day after day. Sometimes, there was a refreshing, but undoubtedly very drying, north-easterly breeze. One Saturday I had arranged with a young doctor at the Mount Vernon hospital near Northwood, whose mother - also a widow - was living in the same guest house near Newnham College as mine, to pick him up in mid-morning and drive him to Cambridge. But I had a strange experience when I woke about seven o'clock. I was paralysed. Having very great difficulty in rousing myself, I decided to have another half-hour's sleep. Then I managed, still with great difficulty, to shave. The effort of getting the razor up to my face was as much as I could manage. Mystified, and exhausted, I went back to bed again and slept another hour and a half. Then, I found the paralysis had gone. So I got up, had a little cereal and coffee, fetched my car, and so left for Cambridge. When I picked him up for what was a lovely drive in my open car, my young doctor friend told me:

"You have evidently just had poliomyelitis with only a minor seizure". The epidemic was rampant at that time, but I had too newly returned to England to have heard, or thought, much about it. There is some virtue in *mild* attacks of potentially serious illnesses for the prospect of immunity which they confer!

As after my canoeing accident in the River Shannon off Foynes Island in 1942, when I could so easily have been carried away by the current, and drowned, I felt a strongly renewed obligation to use my life to make a worth-while contribution to the lives of others and their appreciation of the gift of life.

There were meetings of the Royal Geographical Society in June and November 1947 to hear about the *Balaena* expedition from John Grierson on the work of the air party and from me on the meteorological experience, to discuss the results and lessons to be learnt. Great interest was taken in the air exploits, which were seen as having proved their practicality as a means of whale spotting, and in our meteorological observations, analysis and forecasts, and especially in the geographical discoveries and deductions reported. Old Sir George Simpson, who had been meteorologist on Scott's expedition in 1912-13, came out of his retirement to hear the account. He expressed a proper scepticism about deducing the existence of an Alpine-scale barrier, a great ridge or the ice crest in the heart of the Antarctic continent close to a line from about $80°$ S $80°$ E to near $70°$ S $97°$ to $100°$ E, as a sort of backbone to East Antarctica. He pointed, out quite rightly that my observations were relatively few, and from only one season, and that we knew from northern hemisphere experience how differently the weather systems can be placed, and what different tracks they can follow, from one year to another. But the north-eastern end of that line was right in the longitude sector where we were working most of the season and probably within 500 nautical miles of our ship, a distance of the same order as that from London to the Alps and less than that from London to the mountains of Norway. I was convinced that the extraordinarily repetitive differences distinguishing the wind pattern from one group of longitudes to the next, revealed by ships' observations south of latitudes about $62°$ to $65°$ S, off the coast of Antarctica, not only during the year just past but scattered over all past years for which we had many records, required some great fixed barrier in the interior to explain them. The frequency of southerly winds, counting all cases from S, SE or SW, was under 32% between $75°$ and $95°$ E, but ships reporting between $105°$ and $115°$ E showed over 52%, and the winds reported between $75°$ and $95°$ E appeared generally lighter than in longitudes immediately east and west of that sector: surely a hint of some sort of shelter effect even at that distance from the supposed barrier. Farther east, particularly, the frequencies of southerly winds and of storm force winds were known to increase and near 140° E on the coast, in the territory known as Adélie Land, Mawson's expedition base, Cape Denison, reported winds from S or SSE at nearly 100% frequency; a truly remarkable figure. And gale force was reported at one time or another on nearly every day of the year. Such extraordinary figures are surely only possible with some permanent topographical control. Less than ten years after our expedition, international activities connected with the International Geophysical Year 1957-58, and most specifically the establishment of the Russian Vostok base on the ice cap near 77° S 108° E, at a height of 3,500 m (about 11,500 feet), near a prolongation of our suggested line, confirmed the predicted barrier. Surveys now available show that it rises more than 4,000 metres above sea level between 80° and 83° S at 65° to 81° E and extends to near 70° S 135° E, where it is still almost 3,000m (about 10,000 feet) high. It is indeed the ice cap crest.

I was less successful with my suggestion of a great valley in the interior of Antarctica, channelling the cold air drainage towards Mawson's base on the coast at 140-143° E, which his expedition called "the Home of the Blizzard", although this too gained considerable acceptance at the same Royal Geographical Society meeting. It now seems that the cold air drainage is only very broadly concentrated between the 2,000 to 4,500 metres (8,000 to 15,000 feet) high South Victoria Land mountains along the western edge of the Ross Sea and the "nose" of the ice-cap crest near 135° E. The extreme concentration of the south-easterly and southerly winds near Mawson's base must be attributable in some way to the slopes of the ice sheet and their effects on such dense cold air.

Through the early part of the 1947 northern summer my thoughts dwelt largely on quite another matter, which was to have important consequences for me: the opportunity that seemed to offer of visiting the Faeroe Islands for a summer exploration while communications between Britain and there were still far simpler, as a relic of the British occupation of the islands during the recent war, than they had been before the war. But maybe it was already too late. Some time about 20th July, it dawned on me that the splendid summer of that year was "ticking away" and I was no nearer finding out how I could get to the Faeroes than when I started to think of it as a holiday plan, months earlier. Since the continually cloudless skies extended as far as the Highlands of Scotland, it seemed to be the right year to go there. Old friends and neighbours of my parents, Jock and Meg Elliott, were on holiday with their children near my old haunts in the Cairngorms, staying in Kincraig, a village in upper Speyside close to the mainline railway to Inverness. So I asked them by telephone to find me lodgings near by. And in the middle of the next week I travelled north with the night train and was soon installed in an estate cottage on the other side of the line from the village, staying with a nice young couple, the Smiths. Mr. Smith was a railway worker. I hired a bicycle which made it easy to ride and walk with my friends, the Elliotts, and also to get nearer to the big hills for the start of some more serious climbs.

The following Sunday morning I attended the kirk in the village, a rather temporary-looking, though well kept, wooden building, with the Elliotts (who were not great church-goers when at home in the Hampstead Garden Suburb, but then we all liked the Church of Scotland services, and there is a special atmosphere about a small Highland kirk). Outside, after the service, in the sunshine the gravelled area between the young pines and the fence opposite was occupied by a great scene of chatting and goodbyes among a big party of family and friends. My attention was caught by a nice, fresh-looking girl, apparently in her twenties, who seemed rather younger than the rest. Back at Dunachton cottages, I asked my hosts if they knew who the great party could be. The Smiths had no hesitation in saying: "Oh, that'll be the Milligans. They come up every summer, and a couple of them have houses here. They're great hill climbers. The younger one'll be Moira Milligan. She and her mother usually stay at Feshiebridge post office, over the other side of the valley, where the road comes down from Glen Feshie." If they were great people for the hills, I decided to call at Feshiebridge one day and see if I could join a party for the hills. I did so at the end of the week after several trips with the Elliotts, who were then going back to London. But I only met Mrs. Milligan, Moira's mother, then an elderly widow, who told me the Milligans had mostly gone back to Edinburgh and the South. Her daughter had gone back to her job as a nursing sister in the Royal Waterloo Hospital in London. "She would have loved to go up the hills with you." So there was nothing more to be done about it than to propose meeting her some time in London. Meanwhile, I climbed Braeriach (4,248 feet or 1,296 m) including a long, lonely walk through Glen Einich. The weather was still hot and cloudless, and I remember saying to myself on the climb: "Why do I do this sort of thing." But the whole place is marvellous, with the big hills and the special vegetation in the glens, and on the heights, and here and there a still loch in the bottoms. And I even did another, shorter climb before going south myself the next weekend.

In early September, in London, I briefly met my father's youngest sister, Dorothy, widow (now for the second time) of a Higher Civil Servant. She, it was, who had advised me to go ahead with the Antarctic experience despite my father's death. I paid a short visit to Hereford, reviving old memories, and then I took my mother for a short holiday in the Forest of

Dean. And then, by arrangement, on the 17th, I met Moira Milligan at her hospital near Waterloo. Eight days later, with what would seem nowadays a foolhardy risk on her part, we met again at Hampstead tube station and walked over the top of the hill, and across Hampstead Heath, to call on the Elliotts at their home in the Garden Suburb. Things went very well. We discovered more points of common interest and fond recollections. So, for the first time in my life, I realized I was looking on someone who could be my wife, who I could long enjoy living with, and very soon it came to me that she probably would be my wife. On February 7th of the following year we were married in the old parish church in Corstorphine, Edinburgh. Moira's father had been minister of that church from 1927 until his untimely death in 1940 after an appendicitis operation.

It is strange to think of the number of near-contacts with her I had had in the past. Many times during my residence in eastern Scotland before the war I had been near my future bride's home in Corstorphine and to her regular holiday stamping ground in Speyside and the Cairngorms. I had even bought my first car, a heavily built (by modern standards) Morris Cowley Coupé of 1930 vintage, for which I paid the princely sum of £7, from the Corstorphine Garage in the heart of the parish. I most of all regret not having met her father, the Reverend Oswald Milligan, of whom she was very fond. He was a highly respected figure in the Church of Scotland, convenor of its committee on public worship and devotion, which produced the church's *Book of Common Order*, 1940 edition, the nearest equivalent in that church to the Church of England's "*Book of Common Prayer*" and in similarly splendid language. As it happens, I had heard him preach two or three times in the early nineteen-thirties in Sunday evening broadcast services from Corstorphine on the BBC and, maybe surprisingly, my father had each time expressed his approval and appreciation of the service and the sermon. There was, in fact, one other leading Scottish churchman besides, a friend of Oswald Milligan, whose Sunday evening broadcast services also won my father's approval, the Rev. George MacLeod, at that time Minister of Govan parish church in Glasgow, who in 1938 founded the Iona Community. That community is a missionary group, of several thousand, ministers and laymen, under a rule of spending part of each summer on the island of Iona, where St. Columba established a Christian presence in about AD 563. This group has restored the ruined monastic buildings on Iona and developed fresh religious and social initiatives in the church.

I suspected, and always feared, from the beginning, that it would be hard for me, having been single for so many years, and probably feeling the need of female company more than most, to settle down. I had been so lucky to have many good-looking girls among my friends, but none of them were of the right age and had quite the serious side to their nature and concerns about society that my nature craved for. Moreover, I felt the need for something of the links with northern Europe, where perhaps such attitudes are commoner, that had left their mark on me all through my childhood and earlier youth. It so happened that our most exciting previous experiences had, for both of us, been in Norway. Moira had been at the age of twenty on a long tour with her father and mother to Bergen, and Oslo, and all through the mountains of southern, central, and western Norway in the summer of 1939, just before the war broke out. That adventure had been paid for by the subscriptions of her father's congregation to congratulate and honour him for the award of his doctorate of divinity. So, by the grace of God, Moira had indeed had something of that experience. She also had the serious, as well as the happy, laughing side to her nature, which was just the balance of qualities that I yearned for. Nevertheless, I was for many years always worried that my habituation to frequent flittings, changes of base involving continual changes of female company, would make it hard - and maybe very difficult for me to settle down to one girl and to be true to her. What helped enormously - one could say, crucially - in making it possible for me to master, and set aside, these temptations and worries, was that Moira was such a sweet girl, so trusting, and having been brought up by her loving parents and the old nannie who was still a family friend, that I could not bear to think how shattered she would be if I ever let down her trust.

With so many, and such particular longings, I must have become almost unmarriageable. But when Moira and I met, and it turned out that she had so many of the qualities and inclinations that I was looking for, it seemed a miracle and dreadful to think that I - or should I say we? - had been so close to missing all the blessings that followed. This, then, was the start of our long, and almost unbelievably happy, life together. Small wonder that we have been happy, when our meeting had had this sense of destiny in it and has been such an endless source of thankfulness. Loved by this dearest girl, I have never ceased to bless the days that brought us together. It seems too that her feelings about it have been much the same. Moira, I now know, would come with me to the ends of the Earth and still have the energy and inspiration to do her bit to look after our neighbours and improve the awareness of community and its bonds in British society, indeed in whatever society, we find ourselves living amongst. And, over the years since, we have come to be richly blessed with our more than usually united family, who share most of these aims.

Through that autumn and winter of 1947-48, the last before our marriage, we were able to see each other in London at fairly regular intervals, while I was busy writing the report for the Met. Office of that

Antarctic expedition with the whalers. There was an interruption in the last month before our wedding. In January I went at the invitation of Dr. Bleeker to the Netherlands Meteorological Institute headquarters in De Bilt, near Utrecht in Holland, to work for two weeks with their young meteorologist, Ben Lopez Cardozo, who had done the same job as mine on the Dutch whaling factory ship, *Willem Barendsz*, in the South Atlantic sector. This collaboration enabled us to put together our weather observations from the two nearly opposite sectors of the Far South and other meteorological reports collected after the event by Graham Britton at the Naval Weather Station at Simonstown in South Africa. Once the whaling operations had ended for the season, the commercial interest's demands for secrecy of the fleets' positions were relaxed, and at last most, or all, of the other working ships' observations were becoming freely available for study. We had the opportunity now, and the time, to compile much more informative maps and on them to re-analyse the meteorological situations over the great Southern Ocean and Antarctic waters with enormously fuller data coverage. We concentrated on the eight-day period 18 to 25th March 1947. This was a period of particular interest because of the encroachment of depressions in over the Antarctic continent from both sides and the absence, for the time being, of any room for a polar anticyclone.

It was possible, too, to study the margins of error in the placing of weather systems on my original working charts from the *Balaena* expedition by comparing them with the maps that had all the, much fuller, information now to hand.

During those weeks in Holland I stayed with the kindly Cardozos, themselves a young couple quite newly married. Together, we visited the famous collection of paintings at the Rijks Museum in Amsterdam. Other memories of that time include the rows of old wooden windmills, still then working, along the river and canal banks - and doubtless a valuable contribution to the country's energy economy. And everywhere, it seemed the towns were pervaded by the haunting, acid smell of peat smoke, once a distinctive feature of the atmosphere in Auld Reekie, as Edinburgh used to be familiarly known ("reek", in the Scots tongue, means "smoke" - cf. *Rauch* in German and *rok* in the Scandinavian languages). The peat burning that was still notably prevalent in the older Dutch towns and villages in 1948 may have been a revival of an old practice occasioned by the wartime fuel shortage. With the Nazi occupation so lately ended, and the Soviet menace already threatening, my Dutch hosts expressed their envy of Britain's position "with that sea in between", even if it is narrow.

Both Moira and I, being the only children of our parents, might well have had some fear of losing our identity if we had married into a much bigger family. This did not arise since we were both in the same situation. There were, of course, much bigger numbers of both Lambs and Milligans in the background, cousins and relatives of older generations, but they were not so close as to contribute any reason for us to feel any unease. That Sunday afternoon in August 1947, when I first sighted the Milligans at Kincraig and the folks at Dunachton Cottage told me of them, we could see from the Smiths' abode across the Spey valley, on a lawn in front of the forest, atop the great bank above the water of Loch Insh, a gathering of a goodly number of Moira's thirty-odd cousins, and at least one aunt and an uncle, coming together at the end of their summer visit to the area. Many of them had been staying in the two houses owned by members of the family which were used as the base for their expeditions. It was an impressive scene. Moira's family numbered more than a few lawyers, schoolmasters, doctors and ministers of the kirk, as well as one bishop of the Scottish Episcopal Church. The records of their lives, and their travels, give a range of fascinating insights into Scottish life in recent centuries.

On 7th February 1948, Moira and I were married in her father's old church in Corstorphine by a friend of his, the Rev. William Urquhart, a fellow Aberdonian born four years earlier than Oswald Milligan, who had by then retired from his position as principal of the Scottish Church College in Calcutta to Torphins in Deeside. My mother, then just 70 years old, came by train from the south and my friend Dennis Conolly was best man. There was a goodly gathering of Moira's family, a number of whom lived in Edinburgh, and one of her aged aunts (who at 82 doubtless was not very sure of her balance) was most unfortunately toppled by me backwards into a flower-bed as Moira and I were hustled by confetti-throwers down the front garden path as we left for our honeymoon. Happily the old lady seems to have been unscathed, and helping hands soon had her back on her feet. Moira's cousin, Will Milligan, who not long afterwards became Lord Advocate of Scotland, had made an amusing speech proposing the toast of the guests. The honeymoon began, dauntingly for Moira around the Hebridean seas, with calls in the Outer Isles but for the second half of the time we were cosily accommodated and looked after in her Uncle Fred's cottage in the woods, at Farr, above Loch Insh, where I had first sighted the Milligan family gathering at the end of their summer holidays in August 1947.

Moira's grandfather, the Rev. Dr. William Milligan, who was born right back in 1820, graduated from St. Andrews University in 1839 before studying divinity at St. Andrews and in Edinburgh. He became a Minister of the Church of Scotland in 1843, the year of the great Disruption, when nearly half the clergy set up the Free Church that was to govern itself free of the patronage of wealthy and powerful individuals.(The breach was healed again in 1929 in our own lifetimes.) William Milligan stayed within the Established Church as a notable liberal both in church and politics. He served first at Abercrombie, Fife, but received his own first charge at Cameron, near St. Andrews, in the following year. In 1845 he was given a year's leave of absence for health reasons, but used this to accompany his slightly younger brother, Peter, to study at the University of Halle in Germany, where he became fluent in the language. All through the eighteen-fifties he was Minister of Kilconquhar, near Elie, Fife, where his father had been Minister. In 1860, William was appointed Professor of Biblical Criticism in the University of Aberdeen and given a house in Old Aberdeen where he lived until his retirement in 1893. In 1868 he had been rejected for the Chair of Biblical Criticism in Edinburgh University "because he was too free in ... his inquiry". I have seen a report that in the flitting from Elie, Fife to Aberdeen his goods were moved by horse and cart. Nevertheless, he travelled widely, visited the Holy Land in 1869 as well as Egypt, Syria and Turkey, represented the Scots kirk in the General Assembly of the Presbyterian Church in the USA at Detroit, and collaborated from 1870 to 1880 with Anglican bishops and Free Churchmen in producing the Revised Version of the Bible. In 1879 he spent the summer in charge of the Scots church in Geneva. In 1888-89 he visited Berlin and all the Scandinavian countries on the European mainland. It is a pity we have no information on his modes of travel.

William Milligan's father-in-law, David Macbeth Moir, M.D., (1798-1851) was a very notable character. He was much loved as the local doctor and surgeon in Musselburgh, on the Firth of Forth near Edinburgh. He was also a great wit, a poet, and the author of several books, some written under the pseudonym "Delta" or Δ, a regular contributor to Blackwoods Magazine. He died in July 1851 of pneumonia after going out in an open horse-trap to visit a patient at night in heavy rain. His statue still stands beside the bridge in the centre of Musselburgh. His forebears, Moirs or Moers, seem to have moved south in stages from Shetland, the first ones settling in Aberdeen before the end of the seventeenth century. My Moira was named after these Moirs. One of Delta's uncles, as a member of the town council in Jedburgh, in the Borders, witnessed the first arrival of cigars in the area and Delta wrote an amusing account of it. The scene was a grand dinner at a great house in the neighbourhood, probably in the 1780s or '90s, to which all the members of the council were invited. The owner was a very wealthy representative of the new rich, who had recently arrived in the area after "making a most tremendous sum of money, either by foul means or fair, among the blacks in the East Indies. The council members felt their quiet lives quite disturbed by such a bizarre occasion and had been very nervous about accepting the invitation." It took several debates and some promptings from the bolder spirits among them before deciding. But, in the end, all their old fine clothes were looked out, the stour (dust) brushed off, the white cotton stockings bleached, and so on. And all the folk of the town were at their doors and windows to "witness the council going away up like gentlemen of rank to dine with his lordship". Most had taken a good draught of toddy to give them courage. At the end of the meal, fresh wine decanters and liquors of various colours were laid down before them, and the toasts were drunk to the King, to the harvest, and "Botheration to the French" and "Corny toes and short shoes to the foes of old Scotland". And "his lordship whispered to one of the flunkies to bring in some things. So in brushed a powdered valet with three dishes of twisted black things....." My uncle, to "show that he was not frighted" took one, and without more ado started in with a good bite. Two or three others followed his example and "chewed away like nine-year olds". Instead of the curious-looking thing being sweet, as they expected, they soon recognized the bitter taste of tobacco. Manners forbade them to give up, and his lordship looked on "dumb foundered", while they chewed and chewed and "whammelled them round in their mouths, first in one cheek and then in the other.....Frightened out of his wits at last that he would be the death of the whole council .....he took up one of the cigars ..... and lit it at a candle, and puffed away like a tobacco pipe. My uncle, and the rest, if they were ill before, were worse now. So, when they got out to the open air ..... they grew sicker and sicker..... waggling from side to side like ships in a storm." And his poor Uncle Jamie was "completely back-balled by the council for having led the magistracy wrong".

  A good deal more has become known in recent times about life in previous centuries in these Scottish districts that are dear to us through the writings of prominent travellers who lived hereabout. There is a published diary from around 1800 of a Highland lady, Elizabeth Grant of Rothiemurchus, whose family built the fine house known as The Doune at Inverdruie, just across the River Spey from Aviemore. They also had residences on Charlotte Square in Edinburgh's New Town (just built when they bought it in 1796) and in the Lincoln's Inn area in London. Their regular journeys north to enjoy the summers in Speyside were a tale of slow progress by coach, thirty miles a day on average, and staying the nights in many good inns along the way. Crossing the Firth of Forth at Queen's Ferry took "an hour at the quickest ... sometimes two or three" in an "ugly, dirty, miserable" sailing vessel. The landing at Inverkeithing (on the north side) was as disagreeable as the embarking... "And it took three more days to reach home from Perth". The conditions of early nineteenth century road travel right down south in Herefordshire, according to one report in 1822, were even slower in that there was no road in that county that was normally passable for wheeled traffic between October and April. But at least the Highlands had been pacified by that time. There is a terrible inscription from the previous century on a well beside the road in the Great Glen, which records a blood feud which led to vengeance being taken several centuries after the cause of it. This makes it the more remarkable that, around 1700, the Earl of Cromertie (*sic.*) was an active Fellow of the Royal Society of London, publishing in the Society's *Philosophical Transactions* the results of his own scientific researches on the state of Scotland's mosses (peat bogs) and forests, how trees in exposed areas near the coast in the north-west were dying, and the wind-felled trees were rotting and rather quickly absorbed into the bog. It is recorded that he used the knowledge and understanding gained from his studies to advise his tenants on management of the land. This remarkable man neatly avoided the hazards of travelling through the Highlands at that time by going by sea.

There has been some mystery about the reports by many of the learned travellers of the Earl's time, and indeed of any time between the 1500s and about 1800 or somewhat after, of permanent snow on the tops of the high Cairngorms and of snowbeds that persisted from year to year in dry gullies in the precipitous northern faces of other Highland hills, including Ben Wyvis (summit 3,429ft/1,045m). Other written reports tell of small lochs in shady positions, which bore permanent ice even in the hottest summer weather. One such, recorded in the Physical Transactions of the Royal Society in 1675, was among the hills around Glen Cannich and Strathglass, in the northern Highlands. Some of these details only came to my knowledge in the nineteen-seventies in the course of my work and reading connected with reconstructing and understanding the history of the development of our climate.

A tremendous increase in the southward flow of polar water in the East Greenland Current in the nineteen-sixties - from about 1962 onwards - to about ten times the average volume of water carried in several preceding decades lowered the salinity of the waters about Iceland and ultimately, in 1968 and 1969, brought polar sea ice to near the Faeroe Islands for the first time for eighty years. This awakened scientists to an instability in our climate, which deserved investigation. Delving into the history of this part of the world's ocean with the aid of fisheries records (some of which go back 300 to 400 years) and the records of the occurrence of ice, one discovers evidence of other occasions, when drastic coolings occurred. Indeed, it was in the Iceland-east Atlantic region that the greatest southward shift of the limit of warm, saline Gulf Stream water took place in the ice ages. To return to recent times, in the colder climate periods that set in abruptly around the early 1300s and again from around 1560 onwards, it is clear that there were substantial advances of the colder polar water in both cases. Ice became really prominent on the waters between Greenland and Iceland, being first noticeable around AD 1200 after not having been seen for a very long time. By the mid 1300s the old Viking sailing routes had to be changed and the Atlantic fisheries seem to have been upset, and ships from various European countries were roaming far west, perhaps as far as the Newfoundland Banks seeking the fish. There was a general recovery to warmer conditions, not far different from today's, in the early 1500s, but that was followed by another, sharper deterioration with more serious advances of the Arctic sea ice. The most alarming stage was reached in 1695 when Iceland was surrounded by the ice, and cut off from international shipping for much of the year. The records of the north-east Atlantic cod fisheries are particularly interesting in this connexion, since the cod cannot live in water temperatures below $2°$ C because of kidney failure. The cod fishing failed in that year all around Iceland, in the whole Norwegian Sea, and in the Faeroe Islands. This seems to be a strong indication that these areas had all been over-spread by the cold polar water of low salinity (which meant that the polar water dominated the surface layer). The only exception seemed to be a "pocket" of warmer Atlantic water, with cod in it, evidently cut off in the inner part of Trondheim Fjord. Meanwhile, the cod were almost absent even from the waters about Shetland. Over the whole of the region here described the prevailing ocean water temperatures must have been about $5°$ C or more below modern averages. Hence, the persistent snow reported in those times on the heights in Scotland should be less surprising. We know that in the extreme year 1695 the polar ice was extensive around Iceland. And in one other year in the same period, 1684, a great extent of what seems to have been drift ice from the polar pack was moving fast into the English Channel, where it formed belts several miles wide along the English and French coasts. It was reported sixteen miles wide at the coast of Holland.

    The latest analogous deteriorations were not on the same scale, but did once more bring the ice back to the waters east of Iceland for some weeks in the spring and caused further investigations to be made. Professor L. Mysak in McGill University in Montreal has noted that the flush of colder water in the Greenland Sea in the nineteen-sixties followed a period of much increased rain and snowfall over northern Canada and flooding of the river systems that flow into the Arctic Ocean. And Dr. R.R. Dickson of the Fisheries Laboratory at Lowestoft in eastern England was able to follow the huge area of colder water on the ocean surface as it passed Iceland and out on to the north-west Atlantic, where it performed a long circuit before returning to north of Iceland by about 1982. Generations have been taught to believe that it is thanks to the Atlantic Ocean, with the Gulf Stream and its continuing drift north-east across the ocean, that the British Isles and even the Norwegian coast far up into the Arctic enjoy a climate that is exceptionally warm for the latitude. But we now see that when the currents vary, as they do from time to time, the exceptional warmth is no longer or so firmly guaranteed.

We find that even in the period from about 1870 to now the ocean surface temperatures prevailing in the region of the Faeroe Islands, in the middle of the broadest part of the channel where warm water of Gulf Stream origin passes into the Norwegian Sea on its way to the Arctic, but is sometimes displaced by a branch of the East Greenland Current from the north, have varied by one whole degree Celsius in their sustained five-year averages. That is twice the range by which the air temperatures over England and central Europe have changed. And the realisation that the waters in the broad region between Iceland and the Faeroes in the late seventeenth century, where the cod fishery failed for at least thirty years and was unreliable for much longer than that, must have been generally $5°$ C or more colder than nowadays means that the average temperatures over the northern third to a half of Scotland were probably $2°$ to $2.5°$ C colder than in recent times. Hence, the anomalous reports of snow and ice.

At the beginning of April 1948, with my report for the Meteorological Office on the whaling voyage to the Antarctic completed, I was transferred from the Marine Division and the Harrow Office to become one of the first batch of four young scientists in the new forecasting research division at Dunstable. We were to work just across the narrow roadway from the wartime huts in which central forecasting office was housed.

So, after a few weeks in lodgings on the high ground at Bushey Heath, our married life began in a charming setting, lodging in a cottage on the village green at Whipsnade. Later, Moira succeeded in buying a small semi-detached house at the top of a hill on the edge of Luton. Moira soon discovered that she was pregnant. Being experienced as a nursing sister, and having many family connexions in the medical and nursing professions, she was keen to secure the best possible gynaecological advice and care. So we arranged for her to consult a private obstetrician, Mr. Lloyd, and when the time came, to be admitted to the County Maternity Hospital in Cambridge. This meant that we had to drive from Luton to Cambridge on a cold and slightly foggy November night, when Moira's labour pains began. And so, just nine months and twelve days after our wedding, our first child, Catherine, was born. It also meant that, each day for the next seven days, I drove those forty miles through the fog to Cambridge to visit Moira and the baby. Mostly, it was no worse than a moderate fog through which one could reasonably drive at about 40 m.p.h., but I did have one alarming moment. Once, when I was nearing Cambridge and driving in daylight at 40 m.p.h., I came over a low railway bridge just after a local steam train had passed underneath. The fog having consequently taken extra moisture injected by the engine, suddenly became a dense wall of "wool", not readily distinguished at any distance from the rest of the fog, but there was hardly any visibility within it. One could perhaps see just five to ten yards ahead, but I certainly could not have seen any stationary obstacle on the road, just beyond the bridge. Luckily there was nothing there, but it was a lesson I have never forgotten. Although I had seen the white steam from the engine as it approached the bridge, the tremendous reduction of vision on entering it took me by surprise. All went well, however, and I even managed to finish painting the outside woodwork of our house without falling off the ladder - despite one alarming moment - before bringing Moira and Catherine home to join me.

My job at Dunstable was in the neat, new one-storey red brick building provided for the forecasting research branches, my place being in the longer-range forecasting side concerned with possibilities of foreseeing the weather development over periods up to a month or a season ahead. I was disconcerted when I found myself sent to work on a particular project that I was convinced from the outset was doomed to failure. My brief was to devise a classification of each day's wind and weather pattern over the northern hemisphere viewed as a whole. Various meteorologists in Soviet Russia - Vangeng'em (or Wangenheim), Girs, and Dzerdzeevski, to my knowledge - had over some years done just that, and I had seen some of their work. All their efforts, in my view, fell down in ways that surely had to be expected. On many days, perhaps most days, the weather map of the whole northern hemisphere could only be put into one or another readily recognisable class, or type, if one closed one's eyes to a host of details which did not fit any of the definitions of the class concerned. The only way of avoiding this difficulty would be to have such a vast choice of types that even the inventor could not be relied on to remember them. And, even if one overcame that difficulty by computerising all the possible solutions, the occurrences of any, maybe of most, of the types would be too few for any useful statistics to emerge. The other possible alternative, of ignoring many details, so as to force recognition that two maps were "really" of the same type was obviously unacceptable.

Luckily, just at that time, we received a gift from an enthusiastic amateur researcher, of a classification of fifty years of daily weather maps, from 1898 to 1947, covering just the British Isles and closely surrounding sea areas. This work recognised just seven or eight different patterns, largely defined by wind directions. I did not think the proposed definitions very well chosen, but the whole thing did strike me as much more workable, and potentially realistic, than the over-ambitious Russian projects. As I was seriously afraid that the plan, which had been presented to me as my main task, would turn out to be a "dead duck", it seemed unlikely to serve any useful purpose and would presumably do nothing for my reputation either. It seemed that the situation might be critical for my research career.

I therefore devised a modified classification of the wind circulation patterns covering the British Isles and surrounding seas between latitudes 50° and 60° N and longitudes from 10° W to 2° E in the North Sea. At first, I used just five different directions of the predominant winds (NW was treated separately from the four cardinal directions (N, S, E and W), because north-westerly winds could, depending on their history, bring either warm or cold air - that is to say air that was either warmer or colder than the sea surfaces over which the wind was travelling - producing different weather and different prospects of cyclonic (storm) development accordingly. Besides, the wind direction types, some days clearly were typical of the conditions in the central regions of either anticyclones or cyclones (depressions). Additionally, honesty compelled me to recognise that some days' patterns were properly to be described as unclassifiable - as many as 8 per cent of the days in some years. Later, it was found convenient to recognise the patterns on some days as of hybrid types, because they either shared the characteristics of two, or even three, types or covered quick changes from one type to another, not altogether unlike, type.

This work led me to spend much of my time working on my British Isles classification, and keeping the much larger sheets of paper involved in my appointed task of considering the whole hemisphere out on the top of my table. I felt like a naughty schoolboy, although my age was then 34, playing under the desk while teacher's attention was distracted! But the history of my childhood upbringing had prepared me for such attitudes through living with the constant refrain: "you *are* a naughty boy". Anyway, the justification was that it seemed the only way open to me to produce a classification that would be of any real use. The head of my branch was a nice, but rather strait-laced, man with a very marked reverence for authority, who I was sure could be easily shocked.

As I neither wished to upset him nor let him prevent me reaching worth-while results, I kept what I was really up to dark until I had results that would interest people to show. These results I submitted to higher authority (Dr. R.C. Sutcliffe) a year or so later, when my immediate boss was away on holiday. What finally came out of this is the longest day by day history of wind patterns that exists for any part of the world. It has been much used and is, I am told, one of the one or two most quoted works in the whole world's meteorological research literature. The original period classified was the 112 years of daily weather maps from 1860 to 1971. The classifying was done in a random order of years, to guard against the possibility of any preconceived ideas of what the classification should show, which might otherwise produce a creeping change or bias in the work. When continuing the record in later years, it has not been possible to repeat that precaution, but despite the numbers of other workers who have by now used this record, there has been no suggestion of shifting standards or inconsistency in the long work. (If there has been any "slippage" of that sort, it seems that it must have been slight.) The record has been kept up to date and has also been extended back in history by applying the same classification to numerous fragmentary periods in the past, for which wind, weather and barometric pressure patterns have been reconstructed from original observation reports in research on reported historical climatic episodes of special interest. At the time of writing, the file of reconstructed daily weather maps covers about half the year 1588 (using weather observations made at the time of the famous Spanish Armada storms) and various shorter runs of days known for severe storms, notably in the 1690s and early 1700s, the 1730s,'50s and whole months in the 1790s, as well as sundry years in the early nineteenth century. The full years 1781 to 1786 are included because all the observation reports necessary were readily available in the Meteorological Office library in the bound year-books of the first ever meteorological society, the *Societas Meteorologica Palatina* of Mannheim, which had been founded in 1780 by the prince, Karl Theodor of the Palatinate of the Rhine. This remarkable early society and its pioneering work in organising an international network of observations which stretched from Labrador across Europe to Russia were regrettably victims of Napoleon's military adventures. Also included are shorter runs of days in many other years in the last four centuries, among them the years of the Irish potato famine in the eighteen-forties and earlier times of stress in Scotland in the seventeenth century. The gaps in the record should be filled to complete such a unique archive of the weather and the world's wind circulation, and it should be completed by reconstructing the patterns from actual observations, with instruments, when and where they can be found. There should be no use of theoretical models, which might impose features from modern conditions that did not apply in the times being reconstructed for study. The limited part of the globe surveyed by this study is the only one for which such a long record could be compiled. It would, moreover, in any case be a promising first choice since it is right in the main zone of prevailing westerly winds and yet an area where that prevalence is particularly prone to break down into "blocked" patterns. It is therefore a place where signals of the general condition of the world's circulation may be seen.

The first fifty to a hundred years of the record provided the base for a number of investigations. These showed that long spells of one or another kind of predominant weather, and the wind pattern that produced it, have been commoner at some times of the year than others, notably in July and August, and so rare around other dates, as in late November and just after, that one could think of the year being divided up into "natural seasons". As this implies, there are some periods of a week or two, around which there is a strong tendency for a change from any type of weather that has been the predominant "flavour of the season". The most noticeable of the five or six natural seasons suggested is the high summer: between about 20th June and late August some seventy percent of British summers seem to develop a more or less set character, be it wet or dry, disturbed or serene.

The records also seem to show evidence, which some of our eighteenth and nineteenth century forebears spotted, notably Alexander Buchan (1829-1907) in Scotland and others in Germany and Austria, of a number of shorter-term incidents in the unfolding round of the seasons that tend to occur in many years about the same dates. Thus, our second daughter whose birthday is on 31st May reached the age of twenty-two before she had anything but fine, sunny weather for her day. (She had lived all that time in eastern England.) The last ten days of June have acquired a similarly good reputation in the London area, where the Wimbledon tennis championships have enjoyed generally better weather than most other summer sporting fixtures. Similar considerations have entered into the occasional debates about fixing the date of Easter. Some of these "calendar-bound" tendencies (or "singularities") were first suggested by our forefathers in different parts of Europe even centuries ago; but, if real, they should have been affected by the change of the calendar in 1752. Some of them one must suppose, should be connected with the tendency noticed for long spells of set weather pattern to break down around particular times. Some became recognised in folklore in various parts of Europe, but it seems certain that specification of the dates of such "singularities"(if treated as key features in the seasonal development) should allow an error margin of about three to five days either way. That is, for instance, the only way that the well-known St. Swithin's day (15th July) rule can be held to make sense. This is the rule stating that if it rains on St. Swithin's day it will rain for forty days thereafter. It seems to express a recognition that by about 15th July most English summers have shown what their predominant character will be.

One other of these old rules may be worth mentioning here, particularly as its currency has ranged over a wide range of Europe between the latitudes of the Alps and Scotland. This is the saying, once popular in various languages and dialects, that "if Candlemas day (2nd February) be wet and foul, half the winter was gone at yule" (old Christmas day, or 6th January on the modern calendar), and "if Candlemas day be bright and fair, half the winter's to come and mair" (more). This rule is certainly related to the general condition (health and strength) of the global wind circulation in the particular year. A very marked feature of the seasonal round shown by our archive is the sharp decline in the general prevalence of westerly winds over Europe's middle latitudes from mid-January to mid-February and the equally marked recovery in late June and July. Between those dates in England westerly winds are, on average, only about half as frequent as in the rest of the year. So, if the weather is calm around the beginning of February, the decline of the westerlies may be presumed to be already well under way: and, in that case the way may have been cleared for a persistent frost period to set in. And in the reverse case the more vigorous westerlies may be able to maintain a flow of mild air from the ocean across Britain and Europe.

The main climax in the winter development of westerly winds over the British Isles seems to come about 8th January, though in some winters there is no noteworthy peak of the westerlies after late November-early December. The central England temperature records collected by the late Professor Gordon Manley showed that of the 94 winters between 1872-3 and 1965-6, just 29 (31%) had no westerly day between the 4th and 12th January: among these 27 cases (i.e. about 95%) produced colder than average Januarys and Februarys and about half included a noteworthy frost period in either January or February or both months. However, using such considerations as a basis for forecasting is so close to the late Professor Franz Baur's methods in Germany (see below) that some of his experiences, which we shall relate in this chapter, must serve as a caution.

One must recognise that there were some shrewd observers of nature and sharp intellects in every generation. Some of them produced by-words on the weather that were useful in many, or possibly most, years though they sometimes failed. In these ways our forebears learnt the ways of nature. Besides the useful rules, our old weather lore includes a good deal of rubbish, some of it probably the result of people who had no understanding of the things involved misapplying the lessons from another country - for instance, in the eastern half of Europe - to regions where they have no direct relevance.

Our archive also shows, as do the longest observation records and diaries from particular places, that there have been long spans of years during which the westerly winds showed a substantially higher frequency across Britain and Europe than they have shown in other periods. The intervals during which there have been fewer westerly winds mean that they were commonly replaced by winds from other directions that brought very different weather and different temperatures, with rain or snowfall. Between about 1900 and 1935, as again after about 1988, and also in the eighteen-sixties, the westerly and south-westerly winds have been particularly frequent. On the other hand, the harsher climates of the seventeenth and early and late eighteenth centuries were characterised by much lower frequencies of westerly winds. And the long-term average temperatures in Europe, and even of the world in general, seem to have gone up and down with the frequency of our westerly winds and the general strength of the world's wind circulation. The change-over periods between these conditions have often been notably sharp. We can see an example of this even in sixteenth century (and some earlier) records: diaries from that time indicate an abrupt decline in the frequency of westerly winds over Europe, including the astronomer Tycho Brahe's observations in Denmark and others around the British Isles, to well below twentieth century averages. Correspondingly, a Swiss observer's record from Zürich shows that the proportion of the winter precipitation falling as snow doubled when the years 1563-76 are compared with 1550 to 1562. These are the broadest and simplest tendencies that have come out of this work. They encourage the belief that the world's climate operates as a basically simple system responding to changes in the underlying energy level (or effective[5] energy level) of the wind circulation, though all the details of geography and timing could turn out to be complicated.

The years after 1550 here quoted seem to have marked the onset of the coldest climate period of post-medieval times.

The forecasting research division under Sutcliffe was also engaged from the beginning in an experiment in monthly weather forecasting for the British Isles, using a combination of physical arguments about the heat supply to the atmosphere at each time of year, and the consequent drive of the winds and moisture from the ocean over Europe, with statistical rules

---

[5] This qualification is to remind us that we have been considering only the energy in the main wind streams. There are doubtless variations in the amount of energy in the eddies and the proportion that this represents of the total energy in the wind circulation.

regarding what followed, derived from the past record. The experiment was continued for over twenty years, indeed for much longer than that as the basis for advice offered to client organisations that wanted it for their business. Experience showed that the success rate of the forecasts averaged 50 to 60%. This was not good enough for the average client to feel confidence or find the forecasts useful, even though the method could be regarded as scientifically justified since there were always more ways in which a forecast could be wrong than right about the one and only outcome predicted. The position was different for professional bodies such as the electricity supply industry and ice-cream manufacturers, who have to make assumptions about the coming weather over a few weeks ahead. They are likely to prefer to assume an outcome based on scientific argument, and the forecasts so derived, to be made available to them. Difficulties affecting confidence are also bound to arise through fluctuations in the success rate, as when a run of successful forecasts is followed by a much less successful series.

Some of the seasonal episodes known as singularities are so common in their occurrence that, when they fail, it may be a useful sign of some anomalous influence that will continue to affect the development of the wind circulation in that year. This was a suggestion made, and used, by the German pioneer longer-range weather forecaster, Professor Franz Baur (1887-1977) of Bad Homburg near Frankfurt. Baur was, perhaps, the most successful forecaster of the weather over periods from about five days to a season ahead, using statistical techniques applied to the records of the previous 150 to 200 years, though even his results had embarrassing ups and downs. He noted how the seasonally increasing strength and predominance of the general westerly winds over Europe in autumn, despite the frequency of a period with blocking anticyclones in November, could be expressed in a useful statistical rule: in years when the first ten days of December, and the first half of December, produced average temperatures in Berlin (Potsdam) that were $2.5°$ C or more above the long-term climatic average, with westerly winds, this regime almost without exception continued to dominate the winter, which can reasonably be forecast to be over-all mild or at any rate no colder than normal. This statement was supported by twenty-four out of the twenty-eight cases in his collection. But there were still four exceptions when the winter as a whole turned out cold, and it was this that led to his worst embarrassment.

In October 1941, when Hitler's armies were marching into the heart of European Russia, and his research institute had been taken over by the Luftwaffe High Command since the beginning of the war, Baur was called upon to issue an operational forecast for the winter ahead. He gave his verdict as follows: "In our weather records there have never been more than two severe winters in a row, so the coming winter (1941-42) will be either normal or mild" (translated by me from the abbreviated version of the original published by Professor H. Flohn in 1992[6]). two preceding winters had been notably severe. But by the time this forecast was issued, it was in serious conflict with current indications of how the weather situation was developing. The cold air incursion over Russia in November 1941 was obviously typical of the prelude to other severe winters in the Russian records, going back to 1742 for St. Petersburg. And, as it turned out, the bitter cold continued and intensified until it approached nearly the 200-year extreme values. Such a forecast failure, having very serious military consequences must have been extremely embarrassing, even alarming, for its author, a scientist employed by a totalitarian regime. Hitler's forces suffered historic losses. There is evidence that, at the time when that forecast was issued, Baur's historical temperature reference data were not yet fully assembled, since the complete records include a run of no less than five winters between 1825-26 and 1829-30 with cold or very cold Januarys and Februarys in the areas concerned. Plainly, the position is precarious for any scientist who would forecast the seasonal weather on the basis of statistical tendencies that are not yet fully worked out *and the physical basis of which is not fully known and understood.*

The position of a forecaster using the modern numerical prediction methods, closely based on physical laws, for just five to seven days ahead is immeasurably more secure. The time limitation to only so few days ahead is linked to the fact that the atmosphere uses up its energy supply in about that length of time and alters the distribution of heat sources for further development in the process. Hence, it seems that even the mathematical forecasting method may reach its limit at not much more than a week ahead.

Hopes of developing any longer-term capability may perhaps be realised through the handling of the more persistent ocean temperature patterns and their slower development. This was an approach that the distinguished American meteorologist, Jerome Namias, was working on with indications of a promising outcome in the last twenty or more years of his life, in the nineteen-sixties to eighties. It has also produced very useful results from work by C.K. Folland, D.E. Parker and others in the

---

[6] From H. Flohn's article: "*Meteorologie im Uebergang-Erfahrungen und Erinnerungen (1931-1991)*", Bonn, *Meteorologische Abhandlungen*, Heft 40, published by Dummler 1992, 81pp.

Meteorological Office in connexion with the famine-causing droughts in the Sahel-Ethiopia zone of Africa since 1968.

While I was working in the forecasting research group at Dunstable between 1948 and the end of 1950, Sutcliffe was organising the best mathematicians (notably F.E. Bushby) in the group to develop the routines for producing a forecast for a few days ahead by numerical calculation. It was the first systematic attempt to build on Lewis Fry Richardson's pioneering work around 1920, but now we were into the time when the computers needed were at last appearing and being rapidly developed, with ever increasing power and speeds. This involved the individuals concerned in being detached from time to time to work with the latest machines wherever they were in Britain or America. The essentials of the pre-existing weather situation were expressed by values at a network of points all over the latest wind and weather maps, for many different levels in the atmosphere from the surface up, and used to calculate the situation to be expected at successive intervals of only a fraction of an hour up to several days in the future. The results were explored to see how far ahead realistic results could be obtained: the effective limit appeared to be in the order of five to seven days, as expected. A much valued further development has been the ability to present forward calculations of the expected rainfall where no estimates could be provided before: from this, there seems to be some important skill in detecting the cases where heavy falls of rain or snow are likely, and taking account of the effects of the moisture present improves the estimates of circulation development. In the forecasting research division I was in a small community of colleagues, nearly all of whom had greater mathematical knowledge and skills than I. There was also a very keen and able statistician, James Craddock. As might be expected, we all gained something from each other's knowledge.

It seemed to me that the ideas of circulation development, largely due to Sutcliffe, but also based on statistical lessons from the climatic record, being put to use in the Dunstable monthly forecasts experiment, were of such value that there was a need to explain them in simpler terms, so that a wider range of scientifically minded people could follow them. I set myself the task of re-stating the underlying principles of barometric pressure pattern development in the simplest language in my writings so that areas within the wind-flow where cyclonic or anticyclonic development can be expected should be routinely recognisable. In this, I was closer to Scherhag in Germany than to Sutcliffe, but my approach did not lend itself to computerised forecasting. It was, however, broadly useful in my attempts to understand and reconstruct the patterns of prehistoric and early historical climates, as in the various phases of the last ice age and interglacials, as well as the course of postglacial times. So I made it my business to make some of the contemporary meteorological writings appearing in Dutch, German and the Scandinavian languages accessible to English-speaking meteorologists. This led me to propose to write an English translation of one, or both, of two German language textbooks, which I believed made the basics of our science understandable to a wider range of people than any corresponding English-language books then available. These were Dr. Richard Scherhag's work, published in 1948, on Weather Analysis and Forecasting (*Wetteranalyse und Wetterprognose*, Berlin, Göttingen, Heidelberg (Springer), and S.P. Chromow's work, published in 1940, an introduction to weather analysis and forecasting (*Einführung in die synoptische Wetteranalyse*), which was actually written from the original lecture notes of the Swedish meteorologist, Professor Tor Bergeron, from which the latter presented in Russia in the nineteen-thirties the concepts of the famous Norwegian (Bergen) group of meteorologists under V. Bjerknes, on fronts, airmasses and cyclonic development. I went so far as to get the authors' copyright permission for these proposals but, sad to admit, I never found time to do the work. These proposals did, however, lead me into fruitful contacts with Bergeron, Scherhag and other leading Scandinavian and German meteorologists in the nineteen-fifties and after, and I hope that some of the messages I learnt from them, and intended to convey to English speaking meteorologists was effectively mediated in my own writings.

Opportunities for further contacts arose when I received notice around Christmas 1950 of my transfer to work in the senior weather forecast office of the UK Met. Office at Bad Eilsen, near Hanover, at the headquarters of the RAF in Germany.

CHAPTER 15. MY POSTING TO GERMANY

So, on a dark day in January 1951, leaving my wife and baby Catherine, now just over two years old, on their own in our semi-detached home in the edge of Luton, I set sail with the ferry from Harwich to Hook of Holland and from there on by train to Bückeburg, in the British occupation zone of Germany. This was a gracious little town with many leafy streets and pleasant parks and gardens around its rather fine buildings, which had come through the recent war relatively unscathed. As little as about a hundred years earlier - in another world, or, at least, another existence - it had been the capital of the tiny state of Schaumburg-Lippe, the smallest of the welter of states that together made up Germany in those days, before Bismarck's time. It still had its air of elegance in the nineteen-fifties, and it may be that places which have once been capitals of a country or region, however small, and formerly had some political and cultural importance accordingly, long retain an air of distinction. There are examples in other countries, including the British Isles. Under the post-war occupation regime in Germany, I was soon allocated a good house, a handsome detached house on a short, straight, tree-lined avenue, Lülingstrasse, with a nice garden and a balcony outside the upper floor bedroom (which Moira and I would occupy when the family joined me), partly overhung by the branches of a big cherry tree. The family were to come as soon as we could wind up our affairs in Luton, sell the house, and store the furniture. We were inexperienced, and it did not occur to us that house prices might rise enormously while we lived abroad. The same course of action might have cost us very dear some years later. However, my transfer to Germany was destined to be much shorter than anyone had planned, and we were not too badly caught.

At this stage of my career the dead hand of bureaucracy showed itself. Nineteen days after arriving in Germany, I was promoted to a grade for which there was no vacancy in the British Met. Office establishment in that country. So, I at once put in a plea to the headquarters in London, in writing, to be allowed to forgo the pay of my new grade for a year or two and stay where I was. I was almost surprised when this request was granted. The staff division of the Meteorological Office had long had a reputation for taking little account of individuals' aptitudes, personal commitments, or preferences, and generally putting people where they would be "square pegs in round holes", although I do believe they were much more considerate where illness or other distressing circumstances were involved. As it happened, I had already had some success a year or two before - and possibly made a nuisance of myself - by interfering with the Office's posting plans for me, when they had proposed to transfer me to Trinidad, which

My Posting to Germany

probably most of the staff regarded as a "plum" posting. I had pleaded that that would be too far away from our two widowed mothers. But what was also on my mind was that a couple of years in the widely envied West Indies might be too beguiling and turn out to be the end of me as a serious research meteorologist.

As things turned out, I was to spend a year and a half stationed in Germany. My family came and joined me in March 1951. Poor Moira managed the sale of our house and the removal of our possessions single-handed, her first experience of that business, but not without an alarming incident when one of the removal men lost his footing in the loft. His feet went through the floor and the ceiling below. But luckily he found himself sitting astride a beam and not too much hurt! When the train bringing Moira and Catherine arrived in Bückeburg station about tea-time on March 27th, they had not realised that it was the next station ahead and their things were strewn about all over the compartment, which was otherwise unoccupied. German mainline trains stop for just one and a half minutes at the minor places along the line. So there was need for haste. All went well, and nothing was left behind on the train, but little Catherine - then aged two years and four months - was lifted and popped through the lowered window into my arms as I stood on the platform. She found the re-union a wee bit abrupt, and looked at me very doubtfully after the two and a half months without seeing me!

However, she soon began to take an interest in all around her, was quick to recognise the most commonly seen German (and other) makes of car, and picked up some words of baby German. (Her "Schulgenen Sie bitte" for "Entschuldigen Sie, bitte" (meaning "Excuse me, please") was quite a triumph before the age of three.) And the products of the German car firm DKW (Deutscher Kraftwagen), pronounced Day Kah Vay, which Catherine called "Baker-vade", were obvious favourites with her. Now in her mid-forties she has long been teaching German and French in English schools.

We soon set about getting a new car, with the benefit of the price concession on purchase tax for British government servants working in the occupation regime then in force in Germany. This was a valuable opportunity for us, since our first post-war car had been a disappointment, which "landed us" in eight breakdowns on the road in its first year, all of them involving a tow home. Being of 1947 manufacture, it had probably stood for months in some field, rusting inside its petrol tank or feed-pipes, while the factory waited for some of the parts needed. We went by train to the Rhineland to pick up our new Morris Oxford (1951 model) from the dealer, and that was the end of our motoring troubles. On the drive back we stopped for tea at a tea house which, by somebody's bizarre idea, was in a disused windmill. The sails were still there, and seemed intact, but a row of windows had been put in all round the entrance level, and the teas were served where the machinery had been. For two or three years after that, whenever we drove past a windmill, our small daughter would inquire eagerly: "Is it a tea one?" But we never saw another.

Our transfer to Germany also brought a new interest to the lives of our two lonely, widowed mothers, who were living in Cambridge and Edinburgh. They bravely overcame their fears and their difficulties in getting on with each other, and kept each other company on the long journey by ship and train to visit us in Bückeburg. The visit in September was an undoubted success. It was a good time for the grannies to be re-united with their one grandchild and with the excitement of new experiences. Bückeburg is at the southern edge of the flat north German plain just where the rocky outlines of the wooded hill country, the Weserbergland, begin. So there were picnics in the sunny open land and in the woodland clearings. But the favourite treat for grannies and grandchild was the short journey by train on the electric Kleinbahn (little railway) to Bad Eilsen and then on by another local line to Rinteln, a picturesque old town on the River Weser, where one of the steep-roofed houses was a cafe: after eating two of their delicious cream cakes and drinking a cup of coffee, one did not feel the need of much more food the same day. This treat could be varied by taking another train, a steam train with a great big engine, at the junction near Rinteln to go the few miles east to Hameln (or Hamelin) with its medieval timber and brick buildings, a number of them bearing inscriptions of ancient wisdom and the rat catcher legend in old German. One noted that in the old inscriptions the rat catcher was described by the word "piper", and other words too were nearer to the English forms than modern German. Nevertheless, on the whole it is the German language that has changed less with time, and the strictness with which German preserves its correct forms and grammar is an attraction and surely a virtue of the German language in these times when everything changes so fast that English grammar has almost been lost in a single generation since the war. Italian and Swedish may sound more beautiful, and English may be foremost for adaptability, but German is so clearly spoken by many of its native speakers that it must be the easiest language for foreigners to hear correctly.

One of our favourite areas for walks was in the hills surrounding the Extertal valley just south of Rinteln with its wooded hill-tops and lovely tree colours in autumn sunshine. The old ladies also enjoyed being taken to the broad Steinhude Meer, a reed-girt lake about five miles long, in the edge of the flat land just a little farther east, towards Hanover.

My Posting to Germany

Somehow or other, in spite of the grannies' visit and occasional night duties in the forecasting office at Bad Eilsen, I managed to complete a scientific paper that my old friend, Professor Gordon Manley, had urged upon me during his final term as President of the Royal Meteorological Society, on the "alimentation (ghastly word - Manley's, not mine!) of the Antarctic ice-cap". It was based on my whaling voyage and my observations and daily weather maps which showed that ordinary cyclonic storms from the ocean, with their fronts and cloud systems, did at times invade the Antarctic continent and were on occasion seen to be re-emerging from that frozen land to come out over the ocean again. So there was no need to invent a weird and wonderful theory - such as Hobbs had proposed in the case of Greenland - of how moisture might be directly deposited from greater heights on to the ice-cap to account for its sustenance. Carefully calculated estimates even suggest that the great Antarctic inland ice sheet may be growing at the present time. Ordinary precipitation was clearly occurring by processes analogous to those over terrains with similar relief elsewhere in the world. The paper was duly delivered at Manley's invitation at the four-yearly congress of the International Union for Geodesy and Geophysics in Brussels and was well received. I travelled by train from Germany and returned as soon as possible after the meetings to my family.

We also managed a summer holiday in Norway that year with our new car, by extending a journey that I made in order to visit the famous German marine meteorological institute, the Deutsche Seewarte, in Hamburg. The earliest series of daily weather maps covering the North Atlantic Ocean, which ran from 1873 until the 1890s was produced by collaboration between the Hamburg Seewarte and the Danish Meteorological Institute. The Deutsche Seewarte nurtures a great record of research and practical weather forecasting for the needs of shipping, and the fisheries on the northern seas in particular from early days in the nineteenth century, supported by publication of its own journals which continues today. (It is interesting to recall that until the 1860s the Danish-German frontier had been right at the edge of Hamburg in Altona and to note that, despite that so recent conflict between the authorities of the two nations, this highly respected scientific collaboration was already possible in the 1860s, '70s and '80s.) I subsequently came to know Dr. Martin Rodewald of the Seewarte and some of his colleagues quite well.

Our family holiday in 1951 took us on by road from Hamburg into Denmark and north over the Skagerrak to Oslo, where we bought a tent, and then we turned south and west to the gentle, but rugged scenery of the south coast at Fie (pronounced roughly as Fee-ya in English transliteration), near the outer end of Sandnesfjorden. This choice was recommended by Moira's

My Posting to Germany

cousin Sheila, who had married a Norwegian naval officer, Hans Hostvedt, and had visited us in Germany. We found a very picturesque area for our camp, in a meadow in gently undulating country beside the fjord and put up our tent by permission of the owner of the field in the slowly gathering dusk of an evening of lovely clear beauty - only rather late for wee Catherine - around 10 p.m. About five hours later we were wakened by a very noisy thunderstorm and then, soon, the rain, heavy and persistent. A little later it started to come through the tent as a fine mist! By 8 a.m. - it was a Sunday morning - everything in the tent was getting very damp and we were worried. But when we unlaced the tent door and looked out, there, near us, stood a little red-haired slip of a girl of about twelve years, calling us and saying: "My auntie says you cannot stay there in this weather. She has huts, with rooms, to hire. They are all full just now, but you must come and spread out your things on her sitting room floor. Then, tomorrow, you can move into one of the huts." We accepted with great relief and gratitude. And the girl's auntie Tante Magna, proved as kind as she sounded. She was a big blond lady of probably fifty to sixty years, who reminded me strongly of a very Nordic (but, in fact, Irish Catholic) aunt of Trevor Huddleston's who lived in South America. There was one more chapter in the story of our experiences at that camp site. Tante Magna was deeply shocked by the poor performance of our new tent, made in Norway. And as soon as Monday came, she insisted on expressing her disgust strongly on the telephone to the shop in Oslo where we had bought it. The result was that we got our money back without making any effort ourselves. We packed the tent up, when dry, and sent it back to the firm in Oslo. Meanwhile, we moved into a typical Norwegian "*sommerhus*" (summer house), a one-storey ample sized wooden dwelling with electric cooking facilities and water supply laid on, beside Tante Magna's place, costing at that time the equivalent of £1.40 a week, and enjoyed its nice-looking simple summery garden-type wooden furnishings.

The weather unfortunately continued very broken, and we spent much of our time gathering berries in the woods nearby and using our car to explore the largely wood-built, bright - mostly white with red roofs - little towns along the coast. But we were happy, feeling ourselves among a friendly people with a leisurely pace of life. Over the next three summers after that, having been transferred once again by the Met. Office, this time to the heat of Malta, we often wished ourselves back in the wet woods near Sandnesfjord. We have over the years since had many holidays on the same coast, and that first visit in 1951 remains the only time we had bad weather. It is a favourite coast for old sea captains to retire to, and the older houses are often as charming as the natural scenery. The underlying, meteorological

My Posting to Germany

reason for its good climate, particularly its good summer climate, is that the frequent west and north-west winds shed most of their moisture on the west side of the Norwegian mountains and habitually leave the south-east-facing coast sheltered. Even the sea water often becomes very warm for the latitude when the sun shines for long hours on the rocks and sand. But in really unsettled weather, when the westerly and north-westerly winds are stronger, much colder water may be drawn up from the bottom of the bays.

Back in Germany, we set out in October 1951, driving south-east to south across the middle of the country about two hundred miles to Bad Kissingen for the Tagung, or annual gathering, of the German meteorological society. The whole journey at that date was a marvellous pageant of tree colours. We took with us the senior German meteorologist working in our office staff at Bad Eilsen, Dr. Johannes Enge and his wife, Eva. (His family came from near Hamburg and so it happened that his grandfather came from the Danish side of the old frontier and so had been Danish.) They guided our journey by way of the old roads which took a very interesting route following the Weser valley to the impressive old town of Hannoversch-Münden, with its many-storeyed carved wooden houses from the sixteenth and seventeenth centuries.

The meteorological gatherings were held in the stately assembly hall of the spa at Bad Kissingen. I do not remember the meetings on this occasion for any striking advances in the science of meteorology, but it was a memorable occasion none the less for its review of the useful working practices that had grown up in Germany during the war and after and the ideas underlying them. The gathering was conducted in fine classical style from the live musical opening with Brahms' Academic Overture performed by a small orchestra, followed by the introductory address on his own scientific work on pressure waves in the atmosphere, given by Professor L. Weickmann, an internationally respected father figure of the German meteorology of that time, who had done what he could to make conditions bearable for the Norwegian meteorologists under the Nazi occupation during the war. Moira and I were also kindly included in a party given by Dr. Scherhag, where our sweet natured little three-year old daughter Catherine with her golden hair was much admired. The return journey took us through the smallish university city of Göttingen, very close to the border of the Soviet occupation zone of Germany, the Iron Curtain - just a grim memory now of the endless trench marching across the verdant countryside, scarred with barbed wire, tanks and look-out posts. But Göttingen is also in our memories for Moira's father's studies there in the last century, and our Catherine's much later sojourn there as a student, and not least for the charming little fountain in the city centre with its beautifully wrought iron statue of the little goose girl from the brothers Grimm's story. Sad to say, I also remember Göttingen for a completely sleepless night due to the clatter of traffic over the cobbles echoing along the street due to the house walls, a nuisance to which historic towns on the continent seemed particularly prone and undoubtedly at a new peak in the era of light, almost un-silenced two-stroke motorbikes.

There was one other member of our German meteorological staff that we became friendly with - Dr. Max Ackermann and his wife were older than the Enges and had somewhat indifferent health. They were a childless couple and exponents of the German system of taking their cures twice a year at this or that Bad, a treatment approvable by the German health service arrangements. We continued in occasional correspondence with them for twenty years until Max died.

When I arrived in Germany in January 1951, I had with me an introduction from the Quaker headquarters at Friends House in London to a like-minded, elderly Lutheran pastor, the Rev. Dr. Mensching, whose village church and parish were at Petzen, just outside Bückeburg on the beginning of the plain to the north-west. He had been a missionary in Kamerun (The Cameroons) in West Africa at the time of the First World War. He was a colleague of Dr. Albert Schweizer, who became famous for his music, in particular his organ recitals, in Europe, by which the funds for building the mission hospital there in West Africa had been raised. In the years just before we reached Bückeburg Pastor Mensching had established a work camp, manned by volunteers from many countries, a continually maintained group of young people, which constituted the International Friendship Home. I never managed, however, to go to more than one or two of their meetings. It seemed more appropriate and effective for me to act as a friendly contact, hoping to be a bridge-builder and to exercise a healing influence in contacts with our former enemies in whatever situations arose. We had seven or eight German staff working in the meteorological forecasting office at Bad Eilsen three or four miles east of Bückeburg. The Mensching family had had their troubles with the Nazis and two sons had been lost in the war, believed shot in the back, on the eastern front. We became friends with the whole surviving family. Another son, Fritz Mensching, who was a specialist doctor in the main hospital in Hanover, has visited us and our daughter, Catherine, over the years in England. It was a sad moment when the time came to say goodbye to his father and mother in 1952 when the Met. Office switched me to Malta. The old pastor was fond of our dear Catherine, *"Liebe Cann"*, as he called her, using her own baby-name for herself. We have happy memories of the peaceful summer tea party

My Posting to Germany

with the members of his family in his vicarage garden, surrounded by the big trees, for all the world a scene familiar in England.

To the other side of Bückeburg, the small roads by which the military cars that took me to Bad Eilsen to work went through undulating meadows in which many of the trees hung with mistletoe. Then, a couple of miles farther on, a pleasant ridge loomed higher before us, clothed with pinewoods which neatly hid a mass of earlier small open-cast coal pits, so well hidden in fact that after perhaps a hundred years the woods were pleasant to wander through and, in spring sunshine, one was liable to meet small groups of local school children singing spring folk songs: "The sun shines into the world - do even so - be happy and glad". The pretty village on the top of the ridge, Obernkirchen, is the home of a famous children's (church) choir. There was another good walk, up and along the west-to-east running ridge that rose up right at the end of our road in Bückeburg. This walk led the way up a broad avenue of "acacia"(*Robinia*) trees to the ridge, and farther on some three miles along the ridge, to Bad Eilsen, passing an ornamental tower, the Ida Turm on the highest point, very much like the Leith Hill tower in Surrey, the hill being about the same height. This was a pleasant way to go to work when time permitted.

My place of work, the meteorological forecasting office for the headquarters of the Allied Second Tactical Air Force, was in the (temporarily) adapted upper floor of the Kurhaus of the spa, Bad Eilsen, a comfortable room that extended the length of the upper floor, and looked out of its tall windows straight onto the beautifully laid out Kurpark with its beds of flowers and bushes and trees. We got used to bats - some of them quite big fellows with speckled breasts in the same brown and pale fawn, almost white, colours of thrushes - flying in from the park in the warm summer nights and then back and forth up and down the length of the room above our heads. Our job was to prepare weather forecasts for guidance to the sub-offices on the airfields in the British occupation zone and for the military command's planning. I was lucky that all my time there the situation was quiet, without the extreme tensions and activity that had gone on, for instance, at the time of the Berlin airlift in 1948. I took my share of both day and night duties, always as the one British forecaster on duty with the German staff, which consisted of an assistant forecaster with weather observers also and weather mapping assistants. I never really liked the broken nights schedule with two hours for sleep on a bed between about 2 and 4 a.m. But that meant that one got time for a morning's sleep at home afterwards and then one and a half, or sometimes two and a half, days free. So one had grand opportunities for day-time breaks to explore the town with its shops - there was one really good book shop - and the countryside, as well

My Posting to Germany

as for sight-seeing and social life. There were German language classes provided free of charge for British personnel - soldiers and their families and all sorts of associated workers at the headquarters and outlying establishments. Yet only nine people out of the substantial numbers of British and allied staffs around took advantage of the classes!

There was a better take-up rate at the Scottish dancing classes and the various church activities.

Living in the heart of Europe offers great chances to visit and explore many places and areas that are much farther from Britain or Scandinavia. And my roster of duties, with regularly recurring breaks after night duty, was especially favourable for this when combined with a few days from my annual leave entitlement. In this way, we managed a Whitsun visit to the Alps just across the Austrian border at Ehrwald, where there was a leave centre for Allied Forces. That was a great time for the wild flowers, and we took little Catherine up pleasant paths through the forest and clearings to about the 7,000 foot level on the Zugspitze, the main peak in that part of the range. The gorgeous orange-yellow, big round Trollblumen, which she called "'tatoes", particularly appealed to her. In the two winters that I was in Germany, we took full advantage of the leave-centres for the troops in the Harz mountains at Bad Harzburg and the mountain and village of Hahnenklee, skiing and gaining more experience of the variety of snow conditions and how they change. At the Harzburg centre, for our entertainment, there was a pianist who played Mozart's Turkish March and many other fine works on a grand piano. There were four or five of us British weather forecasters at the Bad Eilsen headquarters, some of whom we were still in touch with long after our return to Britain. One of these was Bill Hogg, who became a king pin in the Agricultural Advisory Service. Another, who was not himself stationed at the HQ but on one of the airfields near, was Michael Hunt - a memorable character, shortish in stature, but with a broad, characteristically beaming face and high complexion and with rather thin, reddish hair, accompanied by a finely twisted Kaiser-Wilhelm, handle-bar moustache in the same colour, a work of art that he was manifestly proud of. Michael and I were to meet again many years later, when he was the first Anglia television weather man in Norwich and I was invited to the University of East Anglia to found the Climatic Research Unit.

But in early June 1952 I left Germany on transfer to the Met. Office in Malta. We travelled by car, an exciting journey across the continent through Switzerland and over the Alps to Italy. But it was an arduous journey for Moira who had suffered a miscarriage only a few weeks before.

CHAPTER 16. AND THE SWITCH TO MALTA

There were other aspects of our transfer to Malta that caused us some misgivings. The journey itself, however, with our car, going over the Alps in June by the St. Gotthard pass was an exciting one. Fleeting memories stay with us of the grand pinewoods in the deep Swiss valleys, the peaks and glaciers, and the Swiss and Italian lakes, all gleaming in the sun. Spring that year in Germany had been a particularly lovely time, with all the blossom arriving together, more than usually close to the same time. But then, as we came down from the St. Gotthard pass, and a night on the mountain at Andermatt, to another world in Italy with its own style of beauty where shutters were kept shut in the afternoons and the red tiled roofs were much less steep than farther north, the sun began to be our enemy. In Malta it was to be the hottest summer for a hundred years. But all down through Italy, day after day our plans for roadside picnic lunches were spoilt by our inability to find any shade. And there were always other children, crowding round the car, surprisingly many of them, appearing from nowhere, always chattering and trying to ask questions, though here the language difficulty limited our efforts. We sailed from Naples to Malta, after a fearsome night in a top floor hotel room, too hot for sleep and, since one could not think of closing windows, one could get no rest from the noises of the city. There were smells too, from a small hand-worked metal foundry, working all night, in a yard five or six floors down, but right below our window. During our drive south across Europe, we had decorated the round nose of our car with little pennant flags of all the countries we had passed through since leaving England. This was not appreciated by the immigration and import control officer on duty at the docks in Malta. Solemn-faced, and doubtless overheated, he abruptly told us: "Take those things off", pointing severely, and with a humourless face, to the flags. We had already felt ourselves under attack in Naples the night before, when looking for a hotel with room for us. Long before we succeeded in our quest, we were scolded by the Naples police for stopping our loaded car outside the door of a hotel: "How do you expect us to guard your property when you stop a laden car on the street outside the doors of the hotel?", they asked us. Clearly, we were being driven to conclude that we were not the right people for the Mediterranean and it was not the right place for us. That evening, after we had installed ourselves in a hotel, recommended by the man I was to replace in the Met. Office in Malta, we went for our first bathe in the sea. Little Catherine unluckily put her foot on the prickles of a sea urchin shell and thereby acquired a septic foot. Fate did not remain so consistently hostile to us in Malta as it seemed on that first day, though there were still more difficulties to come. It was in any case virtually impossible for us to accustom ourselves to the nearly treeless landscape, the over-population, and to living in a garrison community.

After some worrying weeks in a new block of flats in the noisy coast town of Sliema, in which getting a roof over our heads had involved us in signing a lease contract from which we could only extricate ourselves with direct help from the British Air Ministry in London, we found ourselves a pleasant oasis in a palatial old house in the ancient walled city of M'dina on its hill-top position in the centre of the island. M'dina's narrow streets were shady and cool, between the high walls and under the eaves of any houses that had such things. Our house we rented was cool because of its thick stone walls - one of the walls of the kitchen was more than one metre (actually 42 inches) thick - and the ceilings were high: that over our big sitting room was seven metres (23 feet) high. The owners of this mansion had had a bad experience when they leased it to some military family, and from that time on they stipulated that their tenants must be civilians. That put the interesting house within our reach. It had a fine position near the north end of the flat-topped rock which the little city of M'dina was built on. Our upstairs sitting room window looked out over the bastion wall to the sea. And just once each winter, when the air was exceptionally clear, we had a clear view of the snow-covered cone of the Sicilian volcano, Etna, about 130 miles (a little over two hundred kilometres) away to the north.

Our house had a long history, having been according to repute the only house in M'dina that had not fallen down in the great earthquake in 1693. Built on a square plan around a small deeply shaded courtyard, the architecture of its upstairs windows suggested that it had been in Arab ownership, possibly some sort of Muslim monastery, in the time of Arab rule before A.D.1090. Malta is an interesting place because of its archaeology and its history. But the Maltese people did not seem to share the majority British appreciation of its climate. Many of the Maltese, noticeably the traffic police and other officials on duty in the towns, were easily flustered and showed signs of short temper when the weather was hot. One year when the temperature reached 70° F (about 21°C) in really beautiful weather on the last day of February, one of the older Maltese meteorological assistants sighed and said to me, with anxiety written all over his face, "Do you think it is really starting early this year?" In mid June 1952, when we were settling into our quarters in M'dina, the geraniums and the oleanders, even in our garden with its ten or more feet high walls, were withering on their stems, and soon there was no colour left at all. We were already well into that long, hot summer and the fields and houses, everything in the scene, had faded into the dry dog-biscuit colour which one was to live with for several months ahead. At that stage, I am sorry to say, we did not believe my elderly mother-in-law in Edinburgh, who wrote of the great variety of wild flowers that Church of Scotland Sunday School children with their parents in Malta used to press and send back home. We thought her memory was playing tricks with her. But, sure enough, when November came, the flowers began to come back and soon they were there in profusion. They were indeed remarkable for their many varieties. One secluded bay in the west of the island, which we liked for bathing, entailed a walk of half a mile or more down a rocky valley, a sure way of escaping the crowds. The lower end became a ravine between limestone cliffs, where the whole bottom was an impressive natural rock garden, particularly rich in thyme, among other plants. In the evenings, after hot days, it was remarkable to experience the heat retained and still being radiated by the rock slabs long after the sun had gone down.

But even in early October the summer traders near the beaches generally disappeared. Then, as winter approached, the weather became sometimes windy and often surprisingly cold, and unsettled, and showery, some of the showers being very heavy. After the bigger downpours, the streets would run with water like a river between the stone walls of the old city. We were amused, and shocked, by the fraudulence - at least in those days - of some of the tourist literature put out by even internationally known travel firms. We were struck particularly by one folded brochure, headed in inch-high letters "WINTER SUNSHINE IN MALTA", full of pictures of people sprawling in bikinis on sun-drenched beaches. In fact, it was strictly forbidden to wear bikinis in public in Malta, at least in those times. Moreover, on winter days no one would have felt comfortable there in such scanty wear.

The sea is the great amenity in Malta and those in the Navy or with boats of their own have the best opportunities. The neighbouring island of Gozo was a lot quieter than Malta itself. We had tea in the hotel on Gozo in an old fashioned atmosphere, sitting beneath a fine, full-length portrait of Queen Victoria: presumably her death over half a century earlier had been reported, but in the peaceful atmosphere of Gozo, it seemed possible that the inhabitants had not quite believed it. The next year's summer was quite a bit different from 1952. In August 1953, the weather was for some time dominated by what is known to meteorologists as a "cold pool" over the western Mediterranean, and unusually disturbed weather prevailed. This led to us professionally facing an amusing incident. An aircraft flying from Malta to England with an intermediate stop in the south of France, near Marseilles, got into air traffic control difficulties near the French coast and, as is usual in such cases, an internal inquiry was held in the Air Ministry in London. The first that we in Malta knew of this was when a letter came from the appropriate Assistant Director of the Met. Office in London, calling into question our weather forecast because of its stress on the risk of thunderstorms over the western Mediterranean and great cloud development there and over the French coast. The letter from our headquarters in London was quite severe about our astonishing lack of elementary knowledge of the climate of the region. Did we not know that such weather "did not occur over the region in the summer time?" It took me some time to word a reply in sufficiently diplomatic terms, which amounted to saying to the Assistant Director who had written: "please look at the observations"! No more was heard.

    One meteorological investigation that I carried out while in Malta had a curious interest. Because of the island's position in the very middle of the Mediterranean and with very busy air traffic in peace and war, we were well supplied with upper air temperature observations. These were made by radio-sounding balloons released twice daily from a cliff-top site on the south edge of the island's high ground. Each balloon carries automatic registering instruments measuring the air temperature, humidity and barometric pressure (to indicate the height) and transmits their readings at frequent intervals. The temperature measurements frequently showed upper air readings too warm for any buoyancy to develop in the locally heated air over the sun-baked ground of Malta. And yet, on some days when the measurements indicated in this way that no convection was likely to develop over the island, a line of threatening clouds, some building up to considerable heights, did develop: an imposing cloud system" strung out" from WNW to ESE along the axis of the island, which is eighteen miles long and eight miles wide at the widest point (roughly 30 by 13 kilometres). Investigation indicated that this line was the characteristic meeting place of the daytime sea breezes regularly developing over the strongly heated island. This discovery was made easy by the fondness of the islanders and the military authorities for flags (even if unauthorized visiting motorists were not welcome to indulge the same taste!). Regular surveying of the wind directions shown by the flags that were always flying over all parts of the island showed that characteristically a breeze system developed each day with the surface air coming in over the cliffs of both the northern and southern coasts and meeting near the middle of the island. It was over this strip of land in the middle of the island, along its length, that the cloud built up. One need not suppose that the temperatures measured by the radio-sounding balloons were wrong, or really misleading, only that they were not the same as would be found in the locally heated air over the middle strip of the island which was forced upwards by the air streams coming in over the island from the sea over the southern and northern coasts. This forced cloud development did, in fact, produce rain on a number of occasions, although only places underneath the cloud system were aware of it.

We sent our little Catherine to the convent school in M'dina. Her four-year-old rendering of the prayer "Hail, Mary, full of grapes.." was not quite authentic, but she was dressed nicely and she learnt to behave nicely too. We dressed her in beautifully smocked little dresses which have been passed on down in the family since. But we greatly regret that we did not write down the whole prayer with all the quaint passages in Catherine's version of it.

My official responsibilities for inspecting the Met. Office's substations in Libya at Benghazi and Tobruk gave me occasional breaks away from the overcrowded island which was our temporary home. The orderly irrigated landscapes and belts of trees planted, originally by the pre-war Italian colonisers, were a relief after Malta. But as Moira got no such break, we were strongly attracted by a winter family holiday in the Alps. In February-March 1953 we flew to Rome and went on by train to the Austrian village of Steinach am Brenner, to enjoy ten days in the snow while staying in a little Austrian hostelry with its green-tiled Kakkel-ofen heating the dining room and its traditional food. This gave us a real change of atmosphere. Stalin's unlamented death was reported while we were in Steinach. We did some elementary skiing and hauled Catherine back to base on the sledge one afternoon when our expedition was too tiring for her. We frightened ourselves this way, as the wee one certainly got too cold while riding on the sledge. But it was not a very long ride and happily there were no ill effects. A worse incident befell at a stop on the train journey at the foot of the Alps when we were going north to Steinach. We bought hot drinks from a station platform trolley and, while I was paying for them, Catherine seized a plastic mug full of black coffee before I had added any milk. It did not agree with her, but luckily she did not seem to burn her tongue or her mouth and, that time too, nothing worse befell.

When 1954 came, the year in which we were to leave Malta, Moira was once more pregnant and, in May, our second daughter, Kirsten was born at the King George V Hospital for Merchant Seamen. Catherine was no longer our only child, and on the way back to M'dina she told us "I were so assited (excited), I couldn't really think".

On 26th August 1954 we left Malta and headed north, first retracing our sea crossing to Naples. There, our ship was met on the quay by an Italian family bereaved in the war ten years before: now was their opportunity to receive their loved one's mortal remains for burial. The family was out in force, ten or fifteen of them, the women-folk heavily veiled in black and with a great show of surely re-enacted grief that was hard for us to understand however ready we were to sympathize. We drove straight on to Rome as soon as we were free to take the car. In the hotel, in Rome, at breakfast that Sunday morning we needed to buy a litre of milk for the children for our journey, but the bottle would not take the Roman litre!

Any chance of sight-seeing in Rome on this long northbound journey had to be forgone for our two small children's sake. We had seen a little of Rome on our southward journey in 1952, and it seemed possible that some future opportunity would arise in connection with some meteorological conference. Such an opportunity did, in fact, come my way in October 1961 when I was there at a conference under the joint auspices of the World Meteorological Organisation (WMO) and the United Nations Educational, Scientific and Cultural Organisation (UNESCO). But, sadly, my wife could not then be with me, and anyway the delights of the Roman scene were marred by thunderstorm downpours and flooded streets. So in 1954, with a long journey ahead, we drove on north and spent a night, one night only, in Venice. And, thinking that this might well be the one last chance in our lives to see something of that famous city, we did attempt what was meant to be a very brief sight-seeing. But, in the course of it, we ran ourselves into a horrid nightmare, luckily fairly brief, of conscience-stricken panic. We reached Venice early in the evening, garaged our car just outside the city centre and its network of canals, got ourselves into a smallish hotel just across the Grand Canal from St. Marks Square, and were shown a quiet bedroom on the top floor of a pleasant old annexe building, just fifty yards or so round the corner from the hotel. After an early evening meal for the children, we put them to bed in the otherwise unoccupied two or three-storey building, and wandered outside to see what could be seen in the remaining light. Just a few yards away, at the edge of the Grand Canal a motorised tourist boat had tied up with its pilot waiting for passengers. So we asked him did he do just short trips. His reply was rather too much for the one to two hundred words of Italian that were my limit, but I did know the numbers and thought, after repeated questioning, that I understood that his trip would last just twenty minutes or so. We thought we could safely allow ourselves about that long. So we went aboard. Other passengers came and soon we were off. But, after twenty, thirty, forty minutes, there was still no sign of the boat turning round. Our anxiety mounted, but there seemed no other way of getting back to our hotel. At last, after a little more than one hour and a half, we were back where we started, but in a frenzied state. We paid what we owed and ran to the hotel annexe and up the stairs to our room, where we

And the Switch to Malta

found the baby and Catherine undisturbed, in a deep sleep. But it took time for our nerves to unwind.

Next day we returned to our car and proceeded on our way. When we reached the Dolomites, we were really thrilled to find cranberries growing wild in the woods, just as in Norway three years before. We made ourselves a picnic pudding with sugar and milk on the red berries and began to feel ourselves at home once more. In Cortina the new litre of milk that we bought on our night stop would not fit into the Roman milk bottle. We worked our way west to spend the next night in Steinach, and there we found that the Austrian litre of milk would not go into either of the bottles that had come with us from Italy!

As we proceeded farther north, we gradually recognised that the northern side of the Alps and most places we came to from there on, all the way north to Scotland, had been having a particularly wet summer and the persistently wet weather continued into the autumn. There were some beautiful moments of sunshine that made the colourful mural paintings glow on the outside walls of the buildings in Mittenwald, the first town we came to at the foot of the Alps in Germany. We drove on north through south and central Germany, through other old and picturesque - though busy - towns. The countryside nearly everywhere had a decidedly damp look, though we had enough sunshine to see the house walls and their white paint looking clean and fresh. We came to Bückeburg, where we had lived before going to Malta, and we stayed a couple of nights with Flight Lieutenant Jones and his Polish wife and baby, our best friends in the RAF community there. (We were very sad to learn of his young wife's death only two years later and, through successive postings having lost any regular working relationship with the RAF, we then lost touch with the father and child.)

We crossed the North Sea from Hook to Harwich and next stayed some nights with my mother in her guest house in Cambridge, near Newnham College. We then drove on farther north to Edinburgh, to Moira's mother, and then to a hotel in Perth. The next day saw us working our way on northwards along the still in those days very narrow and twisty main road through the central Highlands, burdened as it always was with heavy traffic. Still the woods and countryside about us looked damp and dull under mostly grey skies. We drove over a hump in the road which gave a jolt to the car's roof luggage rack and made it squeak ominously. We had to go on a mile or two before there was a chance to drive the car more or less off the road in an unofficial lay-by, and there we saw that the whole roof-rack and its load had sprung its side clips and lurched forward several inches. But the weight of its heavy load of luggage was keeping it more or less in position. We were able to clip it once more to the two sides of the car's roof, where the rail that was provided for the purpose now held it again. Soon we were on our way again, up by Dalnaspidal and over Drumochter pass. And there, on 17th September, on the moor at the summit of the main road through the central Highlands, the rain turned to sleet. This served to mark the epic nature of our trek across Europe from our friends' farewells in Malta's heatwave just twenty-three days earlier, with the temperature in Valetta still at 96°F (35° to 36° C). But we were happy to be safely home with our two little girls and looking forward to whatever the future might bring. A few miles farther down the road, near Boat of Garten, was the cottage at Mullengarroch, where we were very kindly given house-room for the next several weeks by Moira's old friends, Mr. and Mrs. Jackson (known to us as Uncle James and Auntie Nan).

Mrs. Jackson, a very kind, practical person, had been Moira's Nanny for many years of her childhood. We were now able to spend our after-foreign-service, end-of-tour, accumulated leave with them in Speyside, back among the hills and glens we specially love and where we had first met only seven years before. No great climbs for us now with our little girls, but our days were spent with push-chair and rucksack walking the quiet back-roads across the moor, and over the local Shepherd's Hill, Meall a'Bhuachaille, 810 metres, to give it its older Gaelic name, on the woodland track around the picturesque Loch-an-Eilein , among the red squirrels and other wild-life of the Cairngorm area. The weather had gone into an easier mood, and how much better the cool, clear autumn lights on the hills suited us than the harsh Mediterranean sun or the stickiness of the Scirocco, the warm, humid wind from Africa when it reaches Malta, where it used to deposit beads of water and a general film of water all over the wooden furniture in my office on the airport in Malta!

But the better weather lapsed when, on 25th October, we had to drive south for me to take up my Met. Office duties again, this time in the office at Harrow, and to look for somewhere to live. On the first day of our journey, through the central Highlands the rains resumed so whole-heartedly that we encountered roads that were under an inch or more of water across their whole width even where there was a slope such that the water was running fast. We had to change our route several times at points where the roads were closed by the excessive waters. In that wet, cold season of 1954 it chanced that for about six months, all through the summer half year, in much of Britain the weekends had an even worse record than the rest of the week, a particularly unpleasant record that attracted much attention in the press and in the meteorological journals. But, apart from being one more piece of evidence for a shadowy seven-day cycle, no explanation could be offered. That year achieved the most dismal record since 1922, when the summer was even colder, though many of our past summers have been even wetter, but often with a warmer, more thundery character - 1927, 1931, 1946, 1956 and 1958 to mention but a small selection.

CHAPTER 17. THE CHANCES AND CHANGES OF LIFE IN RESEARCH

I was under notice of posting to the Harrow office. In reality it was in Wealdstone, beside the mainline railway, in a building designed to accommodate a number of headquarters branches of the Met. Office in wartime. The intention was for me to take over charge of the library and the editing of Met. Office publications from George Bull, who was being promoted to the Directorate. But, before the date of my transfer arrived, Bull had a nervous breakdown and was reported to be once more on the road to recovery provided that he could forgo his promotion and have his old job back. This was agreed to, as his health seemed to depend on it. But it left our chiefs with the problem of what to do with me! This was temporarily solved by attaching me as a supernumerary to the climatology division, to deal with climatic inquiries. I am still awaiting my training to be a climatologist! For the meantime I was put in charge of overseas climate inquiries.

The questions being dealt with at that time included guidance for an international engineering firm, preparing to build a dam across the Zambesi river at the Kariba gorge in latitude 16° S in central Africa, and for another firm to dam the Volta river in west Africa (Gold Coast). Another big project concerned a British firm with a contract to build roads in Persia (Iran). The dam builders' inquiries were concerned with the highest flood levels of the rivers which would have to be provided for during the work. The usual practice in dealing with such questions then, as now, was to estimate by standard statistical routines, using whatever past flood and rainfall records were available, the highest flood levels to be expected once in fifty, one hundred or two hundred years and the so-called "return periods" of floods of various heights - i.e. a broad average of the intervals between such occurrences. Commonly such calculations have been based on much shorter runs of observations: the verdict is therefore a fiction, but one that may be justified by its usefulness. In the cases we were concerned with, the weather observations made between 1880 or 1900 and 1950 were used as the basis. Work was begun on the engineering works. But, as things turned out, the flood levels expected once in fifty years occurred every year for three successive years! This alerted everybody involved to the practical problems created by climatic changes and variability. Heavy snowfalls in successive years in parts of south Wales, early in 1978 and again in 1979 and 1982, all exceeding the amounts expected on statistical grounds to occur just once in two hundred years, provided an even more extreme example.

Luckily, the Met. Office seemed to forget about me and my unresolved temporary status for a year or two, unless by chance higher authorities in the organisation had come to hear about a programme of investigations which I had mapped out for myself and liked it enough to consider the results likely to be interesting and possibly formative. Whatever the position really was, it certainly led on to opportunities which I could not have foreseen and which changed the course of my career.

The archives of the Met. Office had at that time lately been reassembled in the sub-basement of the Harrow Met. Office, two floors below the ground, after their war-time evacuation to Stroud in the Gloucestershire hills. They may well have been at that date the richest resource anywhere in the world of past meteorological observations - in a miscellany of publications and hand-written logs from many countries, including the smaller island colonies of the old British empire. Inspection showed that the historical data there should make it possible to reconstruct monthly maps back to the eighteenth century. I decided to concentrate on making barometric pressure maps, since the barometer was early developed to a good instrument in a number of places and it presents fewer problems of exposure to obtain meaningful measurements than either the thermometer or the raingauge. Moreover, the atmospheric pressure measurements should be rather simply related to the pattern and flow of the winds. I therefore decided to reconstruct monthly mean barometric pressure maps covering as much of the world as possible for each January and each July back to 1750. This could be done, once the locations and heights above sea level of each barometer station were known, and the appropriate corrections applied to adjust the pressure values to the equivalent values at sea level and the standard force of gravity at latitude 45°.

The temperature of the instrument also affects the length of the mercury column in the barometer, but this could be allowed for on reasonable assumptions about the temperatures of unheated rooms. (A few cases where the barometer may have been affected by a hot fire near by had to be treated as unreliable barometers.) All this meant a lot of work, and I was happy to be allowed one very reliable girl assistant to do it. The results were indeed encouraging and with about ten points, nearly all in Europe, on the eighteenth century maps, increasing to fifteen or more, including at least two in North America, after about 1780, and a network extending east through the Russian empire right across Asia after 1836. The coverage extended well to the north in Norway and Sweden from the eighteenth century years and to Iceland from about 1820 (though there were also some much earlier observations from these areas in the eighteenth century). The effort involved in the project acted on me as a powerful persuasion in favour of the development of the uniform metric system of units of measurement, since much labour and time had to be spent on correcting, even on finding the definitions of, the older units. Every country seems to have had its feet and inches, but they were nearly all different and were sometimes changed by the introduction of "decimal inches" (ten to a foot) as an early attempt to simplify the system. But what emerged were changes in the atmospheric pressure and wind patterns from year to year and, even when averaged over many years, notable changes in the drive of air from the Atlantic Ocean over Europe and the neighbouring lands. These were variations that made sense with the reported prevailing temperatures and prevalence of rain or snow in winter. The maps were also verified by tests - for instance, drawing the first analysis of maps for the nineteen-twenties and thirties on the basis of only a skeleton coverage of reports such as were available for the seventeen-eighties.

Meanwhile, our family arrangements were in something of a muddle until my posting was settled and we found somewhere to live. Our first solution was to stay at the Bridgewater Arms Hotel at Little Gaddesden, in the Chilterns, at that time one of the old Trust House hotels which I had known before the war. It had a position from which it would be feasible to commute daily to either the Harrow or Dunstable headquarter offices of the Met. Office and to house-hunt for a longer-term arrangement. We soon found ourselves a less expensive berth in a private hotel (the Angleside) in Berkhamsted, where we were allotted a vast attic bedroom from which our small children could hardly disturb the other, mostly elderly, guests. But the house-hunting proved difficult. It was only after viewing 66 houses, and getting into trouble with the authorities for not sending our Catherine (now aged just 6) to school for over two months, that we found ourselves somewhere to live more permanently. We did find a house earlier than that, in Radlett, which would also make it possible to commute to either Harrow or Dunstable. But on the day that the contract on the house we had chosen was to be signed, just two days before Christmas, the widowed lady owner telephoned us to say she had had a nervous breakdown and could not sell. So we had to start house-hunting again as soon as Christmas and New Year were behind us. We had become interested in the schooling offered by a small private school for young children in Radlett, and the lady owner telephoned us in early January to tell us of a big semi-detached house in a quiet road, which a departing parent and his family, were having difficulty in selling. We drove over to see it that day in quiet, but dark grey, January weather, about 4 p.m. We thought after seeing it that it "might do", though nowhere could look really attractive in that weather. We had it surveyed, decided in favour, and were actually able to move in just sixteen days later, on 25th January.

By that time, the weather was turning cold again. I soon set about staining the pleasant wood-block floors of the two main downstairs rooms, which had been fully carpeted, before our furniture came. Unfortunately, the stain turned out to be darker than expected, and I had to set to, washing it all off again while that was still possible, finishing the job after midnight. That involved going many times out and in through the front door, emptying pails just outside down a garden drain. I noticed that it was frosty out there but it was only later that I discovered the thermometer had fallen to -20° C not many miles away. Our floors were saved, but I was far from well for the next few days. The doctor (who was new to us) whom we called was a gloomy-looking woman and inclined towards a gloomy diagnosis. She suspected tuberculosis and arranged for me to go to a TB sanatorium in the outer London area. We were alarmed by the possibility that that could well lead to me catching tuberculosis in the sanatorium. So we sought the help of a new friend, the father of one of Catherine's new school friends, who happened to be in charge of Edgware General Hospital. There I was found to have a touch of pneumonia, which cleared within two days. So I was soon safely back home. I stained the floors satisfactorily soon afterwards, and we registered with another, less gloomy-looking doctor.

The frosts continued frequent in February 1955, and we were able to have some fun on the ice and in the snow. Moreover, even before spring came, we discovered that our new abode in Radlett was a very sunny house, one of the sunniest we ever lived in, before or since. When summer came, we had many meals on the tiled terrace outside the dining room windows. One must conclude that there is some virtue in seeing a house first on a dark, dreary winter afternoon: if it looks like a possible choice under those conditions, it may turn out to be a delightful choice. This one did.

In March 1955 the Royal Meteorological Society announced a prize essay competition in honour of Sir Napier Shaw, who had been the internationally acclaimed Director of the Meteorological Office in the early years of the century, about the time of the First World War. It was one of a series of three such competitions timed around the hundredth anniversary of Shaw's birth. The subject of this essay was to be climatic variation. Obviously, the investigations which I had embarked on in the Met. Office archives fell fairly and squarely within the theme that this competition was aimed at. I had to have a try. The closing date for the essay was to be 1st January 1959. I applied for approval to work on it and I spent many happy hours of intense work in the archives, wonderfully free from interruptions there two floors below the ground, discovering the great wealth of weather observation records that had been collected in the nineteenth and early twentieth centuries, and bound together in handsome volumes here, in the headquarters of the British Meteorological Service, as well as the published collections of other leading countries' weather services, and smaller colonial and island institutions, around the world besides. And there was a good deal of earlier, eighteenth century material as well. I also had the services of an assistant to enter on maps the barometric pressure observation figures - as well as any relevant supplementary information - that my searches produced. And so a series of January and July maps for over 200 years was coming into existence, slowly but encouragingly. But on our family summer holiday in Speyside in 1958 I had to realise that I would fail to reach the target for delivery of the essay unless further help could be obtained. I pleaded my case with my superiors and was generously granted the help of a young meteorologist, Arnold Johnson, from the forecasting division. Between us, we managed to produce all the material required, and a draughtsman provided all the maps and diagrams in the style required for printing. The essay was duly delivered. It did not win the prize. That went to S.K. Runcorn, Professor of Physics at Newcastle University for an essay concerned with far longer-term climatic developments and the course of Earth history, as indicated by the faint remains of earlier magnetization of the rocks, when the Earth's magnetic field must have been in a different orientation - an indication of movement of the poles over many millions of years past. But our essay "on climatic variation and observed changes of the general (wind) circulation" was granted honourable mention as "*proxime accessit*" (runner up) "which seems likely to afford considerable stimulus to the development of the subject".

Our text was deemed too long for publication in the Quarterly Journal of the Royal Meteorological Society, but was welcomed by Dr. C.C. Wallén for printing in two instalments in the journal of the Royal Swedish Geographic Society (*Geografiska Annaler*), which had a record of interest in the subject of the Earth's climate and its development, having previously published distinguished works by A. Defant, Ahlmann, and others. This turned out to be the beginning of an association which finally led to me being awarded the Society's Vega Medal in Stockholm in 1984.

I had been put on the British National Antarctic Research Committee and was still producing research on general polar and Antarctic meteorology and climate. Of particular interest was a paper printed in the Royal Meteorological Society's Quarterly Journal in 1959, which established the much greater strength of the southern hemisphere westerly winds than their northern hemisphere counterparts. With series of daily weather maps, covering both hemispheres complete, becoming available for the first time, the comparative strengths of these two great wind systems could at last be estimated. The results showed that, totalled through the height of the whole atmosphere, the southern system was about one and a half times as strong as the northern one. Much of this may be due to the fact that there is much less friction on the winds over a hemisphere which is 81% ocean. By contrast, the surface of the northern hemisphere is just 60 to 61% ocean. But it is also important that the southern hemisphere is in the midst of an ice age with a heavily glaciated continent at its centre, so that the heat contrast between low and high latitudes is much greater in the south. The southern hemisphere westerlies have a formidable momentum, which acts as a sort of barrier to wind developments from outside. It is associated with centrifugal force that causes the lower latitude winds to intrude across the equator from the south.

This was an exciting time when scientists and the public were discussing the Antarctic discoveries coming from the International Geophysical Year of 1957-58 (a successor to the International Polar Years of 1882-3 and 1932-3) and expeditions which gave us the first complete survey of the Antarctic regions. Until that time there were many blank regions where the basic geography was unknown. It was in 1956 that my finally written up report of the meteorological results of the *Balaena* whaling expedition of 1946-47 was published in the Geophysical Memoirs series of the Meteorological Office - nine years after the expedition's return home. But that was no record for the delays on such works. In 1948 I had received the results of Mawson's Australasian Antarctic Expedition of 1911-14 to review - just published after the lapse of thirty-four years. The release of such treatises, which traditionally print *in extenso* the detail of observations, has generally been treated as of low priority.

Around this time, John Grierson, leader of the air party on the whaling factory ship *Balaena*, on the way from his home in Surrey to an Antarctic meeting in the Scott Polar Research Institute in Cambridge, called at our place in Radlett offering to drive me to the meeting. Normally, it would have been about an hour and a half's drive on the roads of those days. I began to wonder what I was in for when he turned up at our gate in his open sports-model Aston Martin car. He was an aircraft test pilot, and I had no experience of his driving on the roads. It was a beautiful, cloudless and calm, sunny Saturday afternoon in April. The journey took us just three-quarters of an hour, but it was a shattering experience. Everyone was out on the roads enjoying the blithe spring weather. There were young families with perambulators and small children, as well as folk of all ages on foot and on bicycles. The worst moments of that drive came soon after we had passed over a low hill just north of Royston. The lines of the low hedges on either side were very wide apart, but the actual roadway was rather narrow and winding slightly between the broad grass verges, and all down the gentle slope the people and children in care-free mood seemed to be spread everywhere. There was no other traffic in view as far as the blind bend a mile or so on. To my horror, we went down the long slope at 96 to 100 miles an hour. Grierson was clearly driving on the principle of getting past any tricky bits before anything stupid or unfortunate had time to happen. And as we came near the bend, I was astonished at his failure to slacken pace until it became clear that there was a vehicle approaching the bend from the other side. He applied the brakes with impressive results, but we were still doing 54 m.p.h. as we passed about two inches, on my side of the car, from the handle bars of the nearest of two teenage cyclists riding in our direction. My nerves were badly "on edge" still when we reached the lecture room where our meeting was to be held. We got there just fifty minutes before the meeting was due to begin - long enough for us to be in the way of those making the final preparations, but not long enough for us to go away to get a cup of tea! The return journey home that evening was not much better, though the roads were clearer. I remember that as we drove at about 11 p.m. through Knebworth, a big village in Hertfordshire on what was then still the Great North Road, there was an elderly lady crossing the road with her big retriever dog on a lead in the middle of the village, in the 30-mile limit. "Dangerous thing to do at night, madam" Grierson exclaimed, as we braked down from 85 m.p.h. We got home safely, and no one had been killed or injured on our way. But Grierson was visibly tense all the time, for ever adjusting the crease in his trousers and attending to small tickles. I myself could neither sleep at all that night nor remember anything at all of the proceedings of the meeting at any time afterwards!

There were many scientific meetings, concerned with polar and Antarctic affairs and related questions, that I attended in those years. Among them were several at the Royal Meteorological Society and one at the Royal Society in London, as well as the British Association for the Advancement of Science meeting at Sheffield in September 1956 and an international conference in Toronto in 1957, from which I arrived home only just in time to drive my wife to hospital for the birth of our son, Norman, in September 1957. Our choice of the name Norman for him was opposed by my Aunt Dorothy, to whom it implied a French connexion, while for us it had a Norwegian implication. In the end, we established that names were for the parents to choose. (I confess that I have never been of that extreme liberal persuasion that led the parents of an American scientist I once met to christen their son A.E. Smith (or some such name):thirty years later he had never supplied the names that were to follow the initials!) Despite generally, very happy relations with my aunt, and valued encouragement on many occasions, we did have other difficulties with her commanding habits.

About that same period, a young professor from the new Flinders University in Adelaide, Australia, whom none of us had heard of before, turned up on a visit to Britain and made himself known to Aunt Dorothy. When he told her he was professor of mathematics, and asked questions about her father who had been the first professor of mathematics in the original University of Adelaide, she began to regret that a few years earlier she had given me grandfather's silver salver, the rose bowl and candlesticks that had been presented to old Horace Lamb on his eightieth birthday in 1929. Generations of grandpapa's students, going back to the eighteen-eighties had contributed to the subscriptions which produced these elegant gifts. And it had been decided that, as the eldest son of the eldest son, I was the proper person to inherit them. My aunt's second thoughts at this stage were embarrassing in more ways than one. It was startling enough to be asked to give back the honoured gift some years after receiving it. But the worst aspect was that, since the grand things were likely to be so seldom used by the family, we had sold the two silver candlesticks which we did not like - they were in the form of massive Doric columns - in order to buy a handsome modern stainless steel teapot, which we would use regularly and it would be the one item of the legacy to be in regular use! The inscribed silver rose bowl was also far too big to be used more than rarely in a modest household. It was unquestionably handsome and would be much admired. We compromised by handing it over to Professor Potts for Flinders University, where it could be oftener and more appropriately used than by us.

We had had one other, similar experience with my old mother wanting back an item of furniture that she had given us as a wedding present and that she really could not use. So we had to conclude that gifts received from close relations, in their old age, are unsafe: there are seen to be risks in becoming sentimental about such items or ever regarding the gifts as final!

The lectures I wrote had also to be provided in printable form for publication. "Vetting" of original research writings by one's peers i.e. by colleagues who are in a position to comment, and spot any errors, through their knowledge of the subject and related works - necessarily done anonymously - is rightly accepted as providing a needed mark of authenticity, a necessary routine therefore. But the referees' work is not always appreciated by the original authors, who often find it hard to take. Some senior colleagues were so angered by the treatment of their papers that, in at least one case, resignation followed, and the sufferer left to make the rest of his career in North America. Often, of course, valuable suggestions were made by the referees, which improved the piece of writing. But there were also occasions where suspicions of personal motives, and not just scientific judgement, came into the recommendations. One never knew who one's critics were. I also had one amusing, and eye-opening, experience in the case of one of my own main works, on a theme concerned with comparisons and differences between the meteorology of the southern and northern hemispheres. I was required by the referees appointed by the Meteorological Office to remove two whole sections of the paper which were considered unsound. And, when the same paper later came before referees scrutinizing it for publication in the Royal Meteorological Society's journal, I was asked why the contents of the two deleted sections were no longer included. I was then ordered to put them back in, in order that the paper could be published by the Society. And so it was!

In 1959, I received a letter from Professor Harry Godwin of Cambridge University Botany School, author of the classic work on the history of the British flora which had been completed just a few years earlier. He wrote to say that he had learnt with pleasure that the Meteorological Office had at last got somebody working on past climates and inviting me to talk to his group about it, in the series of Quaternary[7] Discussion meetings which he had organised in the botany department. This contact opened up a whole new prospect of work for me, an education in fields of science of great potential interest but of which I knew very little, and an introduction to a world-wide range of many kinds of specialists who had been producing the basic knowledge in those spheres. I was already aware from the writings of my Met. Office predecessors, C.E.P. Brooks and others, that botanists' studies of past pollen deposits preserved in the soil and subsoil, peat and old lake-bed deposits, even also ocean-bed deposits, and larger relics of ancient vegetation, showed evidence of former climates that differed from today's. Indeed most of what was then known of former climates had come from these studies, since the changes of prevailing vegetation corresponded to changes of the climates that nurtured the vegetation. Some of this history had been tentatively dated by counting the year-layers, or "varves" in lake-bed deposits in Sweden back to the end of the last ice-age. The time scale was rapidly becoming more certain with the

---

[7] The Quaternary is the most recent section of geological time, roughly the last one to two million years.

introduction of radiocarbon[8] dating and related techniques that first became available about 1950.

Godwin had long been in touch with many of the leading workers in the various fields of research all over the world, which were contributing to our fast growing knowledge of past climates. Representatives were continually turning up at the Cambridge meetings to lead the Quaternary discussions. I was also enabled to attend international conferences on radiocarbon and other radioactive isotope techniques, some of which had yet other implications. Changes in the rate of production of radioactive carbon in the atmosphere at various times could be detected from the errors they caused in, for instance, radiocarbon dating of objects of known age. These implied changes in the sun and its output which changed the production of radiocarbon and might be a cause of climate change. And there were various other indications of past environments. One learnt of tree remains on the heights of central and northern Europe, including Scotland, at heights where trees no longer grow, of beetle faunas pointing to great changes of climate that must have taken place at different stages during the last glaciation, and of forests which grew on what is now the bed of the Atlantic Ocean, off Argentina, when the ocean level was lowered by all the ice deposited on land. There was much discussion too of the problems of interpreting the evidence found in old lake-beds and shoreline deposits because of the "reworking" that has taken place down the ages of silt and other deposits disturbed by currents and by slumping on the slopes down to the greater ocean deeps. Besides all these, changes of prevailing summer and winter temperatures and of ocean temperatures inevitably registered by indicated changes in the vegetation, fish species, insects, birds and animal remains were reported at these meetings. And these findings led into human archaeology as well. Altogether, my links with the Cambridge botanists and their contacts in the other life sciences in British, Scandinavian and other European, as well as the Irish and North American pioneers of pollen analysis and many other branches of study of the past, were a greatly enriching experience which excited my awareness of many aspects of life and scientific discovery as well as of climatology. This really opened up a second career for me.

As my thinking on the causes of climatic change developed - beyond the explanatory and unifying insights gained from the atmospheric and ocean circulations' mechanism on which I had so far largely

---

[8] Radiocarbon dating is described and explained in more detail in my other books.

concentrated - I was increasingly driven to attempt a thorough assessment of the effects of volcanic explosions and their products in the atmosphere. The first step had to be to build up an adequate data basis. This meant collecting in what time I could spare factual observations and theoretical works covering as many great volcanic eruptions as possible. This seemed one of the most promising, and at least in Britain neglected, likely causal influences involved in recorded climatic changes. So, with that aim, I read widely and was particularly impressed by the writings of W.J. Humphreys in the United States on the physics of the problem, involving particles (one should add also emitted gases) floating in the air in his text-book *"Physics of the Air"*, published in 1940 (McGraw Hill), by the great *"Report of the Krakatoa Committee of the Royal Society"*, published in 1888 (Harrison and Trübner, London), following the tremendous eruption in the East Indies in 1883 and its world-wide repercussions, and Karl Sapper's assessments of all known dust-producing volcanic eruptions all over the world from the year 1500 to the early twentieth century, published in two instalments: (i) Sapper (1917), Beiträge zur Geographie der tätigen Vulkane, in *Zeitschrift für Vulkankunde*, Vol.3, pp.65-197, Berlin, and (ii)Sapper (1927) *Vulkankunde*, Stuttgart (Engelhorn Verlag). There were also very valuable indications of how the effects of volcanic dust and aerosol might work in an article by H. Wexler (1956) on Variations in insolation, general circulation and climate in the journal *Tellus*, Stockholm 1956, pp.480-494 and, just then beginning to be explored, probable traces of the actual working of volcanic (sulphate) acid material found in the year-layers in Greenland's and other ice-sheets by C.U. Hammer and others in W. Dansgaard's Copenhagen laboratory, published in *Nature*, 270, pp.482-6, 1977. There seemed also to be greater possibilities opening up of understanding the action of particular volcanic eruptions on weather and climate from all the new knowledge of the world-wide wind circulation up to great heights, traced by the spread of the atomic bomb test explosion products in the nineteen-fifties and early sixties. (Not an argument for atomic weapons but a case for taking some useful advantage of information available as a consequence of their manufacture!)

I used these studies to write a thesis developing an assessment of the world's volcanic eruptions since the year AD 1500 in terms of the loading of the atmosphere with eruption products and their spread over the Earth by the winds - the ultimate spread should depend largely on the latitude in which the eruption took place - and the indicated "lifetime" (duration) of significant amounts of the suspended materials in the atmosphere. For comparison, some attention was also paid to a few earlier eruptions for which there was good information, such as documented information from Iceland and the Mediterranean and deductive data from

Alaska, Greenland, Kamchatka, Antarctica, and elsewhere. My writing on these studies of the effects of volcanic explosions was quite promptly approved for publication in the Met. Office's series of Geophysical Memoirs, but I myself was not happy for it to go forward without any "vetting" by any expert in volcano science. Furthermore, there was an awkward difficulty: I did not know anyone with the necessary expertise, apart from having had some contacts by correspondence with the well-known Icelander, Dr. Sigurður Thorarinsson, who had however expressed himself favourably on my ideas. I looked up the membership of the British National Volcanological Committee in the Royal Society yearbook, and wrote to its chairman, Professor L.R. Wager of Oxford, and asked if he would be so kind as to look over the work for me. He was very willing and asked me to take the manuscript over to his department in Oxford University. This I did, and he passed it to a Royal Society research student, Dr. P.E. Baker, in his department. Within a couple of weeks, Wager invited me to visit him again. He handed the paper back to me, saying only that it was a splendid work and should be published by the Royal Society. But when I reported this to the Meteorological Office, my superiors' support faded away. In October 1966 Dr. R.C. Sutcliffe went further and told me he thought my paper unsound, based on arguing in a circle, and would be better not published, though he later agreed that this was not so. Very sadly, and surprisingly, three weeks later, the youthful-looking Professor Wager suddenly died. This left me in a very awkward and unsatisfactory position, because, after Wager's high opinion of the work, I was not much inclined to accept any lesser publication channel. Happily, the young Dr. P.E. Baker and the department secretary in Oxford gave me what support they could. But the impasse remained, and my manuscript returned to my office window-sill, where it lay for the next five, or more years, until nature took a hand and solved the problem for me, when a volcanic eruption - admittedly a very small volcanic eruption - took place on British soil (!), in December 1967, on Deception Island, in the Antarctic. This caused the British National Volcanological Committee to meet and the chairman asked: "What British work on this subject have we going on that could be published?" Dr. Baker, who was now a member of the Committee, told them of my paper on "*Volcanic dust in the atmosphere.... A chronology and an assessment of its meteorological significance.*" The chairman wanted at once to know why it had not been published and "Why cannot we publish it?" So the high-ups in the Met. Office reversed their opinion of the work once again, and I was commanded to produce the manuscript for the Royal Society to publish! In 1970 it was duly published in the "*Philosophical Transactions*" of the Society.

The Chances and Changes of Life in Research

My investigations had shown beyond reasonable doubt that great volcanic eruptions do affect the weather and climate for up to several years afterwards, while suspended materials - not only the fine dust, but minute aerosol droplets and even gases - thrown up high into the atmosphere by the blast are still present. The study also showed that it takes some months for the ejected materials to spread round the world and, in some cases, over the whole Earth. It is the greatest explosions from volcanoes in the low latitudes between about 30° N and 30° S that most regularly yield eruption products that reach the atmosphere over all parts of the world. There has been a great deal of research by other workers. The first suggestion of an effect on climate goes back at least to Benjamin Franklin, when as the first ambassador abroad (in Paris) of the new United States of America in 1783-4, he tentatively attributed that long winter to the great eruptions in, and near, Iceland in 1783 and the "universal spread" of "a dry fog .... over all Europe and great parts of the United States of America .... The fog was of a permanent nature .... the rays of the sun seemed to have little effect towards dissipating it .... They were indeed rendered so faint that, when collected in the focus of a burning glass, they would scarcely kindle brown paper .... Of course, their summer effect in heating the Earth was exceedingly diminished ....Hence, perhaps, the winter of 1783-4 was more severe than any that happened for many years ...." In fact, my work indicated that there was some diversity in the nature of such effects from eruption to eruption. There was more regular evidence of cumulative effects when eruptions came in succession, not many years apart. The most regular effects were a weakening of the strength of the general (i.e. global) wind circulation for about the first year after a great eruption which spread matter in the atmosphere over the whole Earth. That could result in either more southerly or more northerly winds over any one area, with, for instance, widely differing outcomes in the winters experienced in Europe. Also, eruptions in middle or high latitudes, which (owing to a tendency to poleward drift in the stratosphere) spread their veils only over the higher latitudes of one hemisphere, tend rather to strengthen the wind circulation because they increase the heating difference between different latitudes. The most regular effects of eruptions seem to be in summer, when there is a tendency for the North Atlantic storm activity to be shifted somewhat south, towards Europe. Hints were found of a like effect over the Far East and the Pacific. In fact, all the wretchedest summers of the last four centuries in Britain and Japan, perhaps also in eastern North America, seem to have been in years when there were great volcanic "dust veils" present.

Later work, in the nineteen-seventies, in the United States by Dr. Steve Schneider has seemed to confirm that loading of the high atmosphere with substances thrown up by volcanic eruptions is one of the most disturbing influences on climate. He found that the changes of world average temperatures from about 1875 to 1975 could be quite well simulated by an equation using only three indicator figures, (i) an index of the amount of recently injected volcanic materials, (ii) an index of the increase of carbon dioxide due to human activities, and (iii) an index of the sun's output. Of these, the last seemed least important. Nevertheless, later work and experience suggests caution and modifications. Studies of much longer-term climate history suggest that longer-term solar output variations may be much more influential than Schneider's study indicated. And it is impossible to go into the details of volcanic eruption accounts without noting the great diversity of different volcanoes and different eruptions. The magnitudes of different volcanic explosions, and resulting differences of height to which the products are thrown - in some cases, there are, too, notable differences in the subsequent diffusion of the products, occasionally to far greater heights - and wide variation of the sizes and chemical nature of the particles and aerosol emissions must surely be always liable to defeat our quest for regularities and rules of thumb regarding the effects.

Many since Benjamin Franklin, notably Defant, Humphreys, Wexler, and others besides myself, have regarded the variations of volcanic activity as among the more important causes of climate's variations. Indeed, the warmth of the early twentieth century was very widely attributed to the occurrence of several decades of quiescence of the world's volcanoes, particularly in the northern hemisphere, and consequent progressive clearing of the atmosphere. It was obvious, even by 1960 that this was a subject that demanded the attention of investigators. But the global cooling, which set in about 1940, and lasted more or less until some time after 1972 and in the higher northern latitudes more or less until 1987, has not been convincingly explained.

Among the most serious gaps in our present understanding are the rapidity of some climatic changes and the timing and causation of some, seemingly abrupt, changes of climatic trend, including this recent cooling and the sudden return to warmth affecting the northern hemisphere since 1987-1988 (at latest). It is not sufficient to say that this is the "greenhouse effect", the warming associated with the increasing carbon dioxide etc. in the atmosphere, which we have been expecting, unless we can say why it did not show in other, recent periods of years over large parts of the northern hemisphere. We can grant that there were some parts of the world where the "greenhouse" warming did seem to be taking effect during those years: in particular, from some time around 1950 there was notable warming in New Zealand and over parts of the Antarctic. But in these areas too the sudden onset needs explaining, and they have not continued in the forefront of the warming tendency.

Such step-like progress, now here and now there, seems to have characterized a number of past climatic warming and cooling episodes. Part of the explanation may lie in preferred wave-lengths in the flow of the upper westerly winds around each hemisphere combined with some degree of "anchoring" of these features by a stubbornly warm or cold surface in one sector or another, or by a mountain barrier interfering with the wind-flow.

There is also still a need for more information about the past history of solar disturbance and for much better understanding of its relationship to climatic variations and the history of the global wind circulation. Research and understanding of any relationships between disturbances of the sun and weather and climate has had a strange history. Since the sun is the source of all the energy that warms the Earth and drives the winds and ocean currents, it was natural that the first thoughts about the causes of changes of climate should be to look for changes in the sun's output. Some prominent meteorologists in the early part of the twentieth century ran into great embarrassment through supposing that this must be an altogether simpler question than it turned out to be. This soon led to a conspicuous failure of forecast. There had seemed to be just the sort of straightforward relationship between the changes of level of the great Lake Victoria in eastern equatorial Africa - a water body of comparable extent to the southern North Sea - as everybody concerned had expected to find in this field of study. The level had gone up and down by about one metre, neatly in parallel with the roughly 11-year variation in the number of sunspots, the lake being highest when sunspots were most numerous and lowest when their number was low. This behaviour was noted during the first three decades for which lake-level gauge measurements (which started in 1896) were available. Effectively, the supposed relationship was working from the known high level of the lake from 1893 onwards. And on this basis, in the nineteen-twenties at least two of the most highly respected figures in meteorology committed themselves to a forecast that the lake would be low again around the next sunspot minimum, expected between about 1932 and 1935. However, the relationship manifestly changed and the lake rose to another high level in about those very years. Looking back from that experience, statisticians would tell us that it had been a much too simplistic - or just a premature - basis for any forecast, since the argument rested on observation of barely three sunspot cycles. The result however was that the subject of sunspots and weather relationships fell into disrepute, especially among British meteorologists who witnessed the discomfiture of some of their most respected superiors. The subject became taboo. Even as late as the nineteen-sixties, despite the successes that Franz Baur obtained with a judicious use of some other statistically derived rules for sunspots and weather relationships, the subject was widely regarded as one to avoid. For a young researcher to entertain any statement of sun-weather relationships was to brand oneself as a crank.

The American meteorologist, H.C. Willett, was the first, writing as early as 1949-50, effectively to revive some interest in the topic, pointing to indications of longer-term changes in the world-wide patterns of the wind circulation, which seemed to be in phase with the longer sunspot cycles of about 22 and 80 to 100 years in length. Willett had some success with forecasts based on this. Such studies have only begun to be possible since indications of long-term changes of climate from the "proxy records" in pollen deposits, the ice-layers in the Greenland and other ice-caps, and isotope analyses of ocean-bed deposits have extended our knowledge of the course of past climate far beyond the invention of the first meteorological instruments.

From the now extended past climate record Dr. Jack Eddy in the United States has focused attention on some remarkable periods of unusually quiescent behaviour of the sun. The two most recent, and very marked, quiet sun periods, with almost no sunspot activity, are known as the "Spörer minimum" and the "Maunder minimum", each lasting some decades, from about AD 1400 to 1500 or 1510 and from 1645 to 1715 respectively. Both were times of extreme, especially cold, climatic anomalies and exceptional year-to-year variations.

CHAPTER 18. FAMILY RESPONSIBILITIES AND CHANGING PROSPECTS

By 1959-60, it was time to be thinking of another flitting. A house-move was looming ahead because of the Met. Office's plans to re-unite all the headquarters branches that had been dispersed during the 1939-45 war to sites in central London, Harrow, Stanmore and Dunstable into a single new building for which a site in Bracknell, Berkshire had been chosen. And this time our plans were complicated by the needs of our family. It would be necessary to re-house our two widowed mothers, both about 80 years old, somewhere in a region where they had never lived before. Some difficulties over that could be expected, and sensitivity would be needed. By this time also the choice of schooling for our growing children was beginning to take on a more serious aspect. Our eldest, Catherine, had her eleventh birthday in November 1959 and would soon be going on to senior school. We held firmly to the principle of using the state schooling system, so as not to contribute to the sadly disruptive effects on English society of having rival schooling systems. I had previously hoped that by this time we might be living in Scotland, so that our children could go through one of the good local academies. In the event, by going to live in Guildford our children were able to profit from very good schools.

The Met. Office was due to take over its new offices in Bracknell in 1961, seventeen miles from Guildford, and, as it was quite clear that the two grannies would be "fish out of water", in Bracknell New Town, we felt obliged to start working on house-hunting early. There was also some possible difficulty over the unusual nature of the population of Bracknell, where in the early years of the new town, a high proportion of the professional population were to be meteorologists. I have always tended to like my fellow meteorologists though a tendency for their conversation to be rather limited is perhaps a characteristic.

So, to be as well prepared as possible, I enquired of the Ministry of Defence finance branch (which would be covering the official costs of the move) whether it would be permissible for us to move early, if and when we found an acceptable solution to the family problems. I therefore went to see Miss Church, the head of that branch, but she clearly misinterpreted my motives or was, at least, very suspicious about them. She seemed at once to suspect that I was trying to invent some advantageous fiddle with the financial regulations governing any allowances that might apply. She warned me sternly not to take any action before her branch had made its arrangements with the Treasury concerning the unusually big move of staff that would be involved with the whole Meteorological Office headquarters. That was understandable, but it amounted to an awkward constraint on the choices by which we might hope to solve our problems, especially for the older members of our family. In the end, having found a house we liked on a quiet road, high up on the hill, in the inner part of Guildford, and explored what seemed to be the best places for our old mothers, we made our move in September 1960 and got our daughter, Catherine, into Godalming County Grammar School just in time to start her secondary schooling there. Both our daughters ultimately had the whole of their secondary schooling there in Godalming. And they were probably as happy as one can hope to be in those stressful, hard-working teen-age years. Each in turn became head girl, and they won themselves places in the University of East Anglia, Norwich and in Newnham College, Cambridge respectively. We had had to go through some anxiety created by the local pride of the Surrey county education authority, which would not accept that our elder daughter, who had been passed by the Hertfordshire education authority for a place in Watford Grammar School - which they considered their best school, rivalled in England only by Cheltenham Ladies College - was up to the standard expected for Surrey's Guildford County Grammar School. We protested, but later decided on experience that our girls were probably better off in the Godalming school.

Visits to the homes available for the elderly in the private enterprise sector in the area around Guildford, and as far as Dorking, opened our eyes to some cases of gross exploitation of those old people who evidently had no alert younger relatives to stand up for them. We did, however, succeed in finding houses that were beyond reproach and where our two old ladies could expect to be reasonably happy. An odd twist to this story was that, some nine or ten years later, after a visit to my mother, we found ourselves briefly visiting Miss Church, who had become a resident in the same home. Evidently, there were by then no lingering suspicions or ill feeling.

In May 1960, after giving an invited talk on the BBC about our increasing knowledge of past weather history and its apparent effects seen in human history, I was invited by Mr. Peter Wait of the directorate of Methuens to write a book on the subject. I told him that I would very much like to do it, but I had long plans ahead for further research and did not feel that at that time in 1960 I yet knew enough to write the book he wanted. That was my introduction to a long and very happy relationship with him and his firm. And, in the first years of it, both sides showed remarkable trust in each other, especially as my ability to start on the project was soon unforeseeably delayed by a personal request I received from my boss, Sir Graham Sutton, the Director General of the Met. Office, that I should write

Family Responsibilities and Changing Prospects

a new edition of old C.E.P. Brooks's book, *"The English Climate"*, to be published by the English Universities Press of London. I did so, and it took me most of the next year, including a lot of my spare time in the evenings. It also put my family to some privations, since our dining room had to serve as my workroom/study and was often cluttered with the papers and references I was using. That book appeared in print in 1964. Sutton was pleased, but I probably made a commercial mistake in adopting the Celsius (or Centigrade) scale at that date. The Met. Office had introduced the "new" scale in 1960, but the British public was clearly not ready for the change, and my use of it probably affected sales of the book.

My contacts in Methuens gave me a nice lunch in a restaurant on Fleet Street about once a year to inquire how things were going on, and I incautiously began to write chapters for the proposed book without having first fixed up a contract. By some time in 1964 it was clear to me that several more years would have to pass before I could provide Methuens with the script of the book that Peter Wait had asked for. So I suggested that it would be possible to produce an interesting book, with a lot of new material for quite a wide public by simply gathering together six or eight of my recently published papers and essays on the changing climate. They did this, and the book ran to a second printing. It was the most labour-saving way and speediest solution to producing a new book that I ever tried, though it did suffer from one very obvious defect. Some of the maps and diagrams were repeated on different pages. But the field which the book covered was so new, and excited so much interest, that there was almost no adverse comment on the duplication. I was told the book sold very well.

It has been a great happiness to me, releasing long frustrated energies, that over about the second half of my career circumstances led me - even required me - to return to my interest in history that was abruptly cut short by the specialization in science that was forced upon me in the middle of my schooling in 1928, around my fifteenth birthday. What has been of importance is not the history of battles and great national leaders - which so many regard as the stuff of history - but of the changes in how people lived, and the stresses and struggles that circumstances imposed upon them.

The historical research which I embarked upon had the unusual feature that it made use of the physical laws that are the scientific basis of modern meteorology to recapture the patterns of wind flow over the Earth and the prevailing weather patterns in each epoch which the wind flow generated. This made it possible to construct the probable global patterns of winds and weather from whatever fragmentary reports and scattered evidence were available. This effort promised to serve the cause of climatic science as well as history. It also taught me many things.

Family Responsibilities and Changing Prospects

The jungle of older scales of units of measurement, which all had to be converted to uniformity before one could use them on a single map, was a persuasive argument for the value of adopting internationally one single set of universal units that should be everywhere the same. More or less every country had at some earlier time used inches and feet and miles, but they were all different. The mile had even had different lengths at one time in different parts of England. Regrettably, the same tendency as in the past to develop confusion still exists today - and even among scientists - who, when new items are discovered that need measurement for study, like to invent new units and, in the same way as our politicians, insist on using only their own versions.

It was an exciting, but excessively busy, time for me. The vagaries of climate in the nineteen-fifties and sixties were more noticeable than for many decades past. And there were, as yet, hardly any other people or research centres anywhere in the world focusing attention on changes of climate and their causes. I was overwhelmed by the volume of inquiries coming in by post and telephone and by the requests for lectures and articles for publication. It was in the midst of this, on an afternoon when I did not know how to get through the work on my desk, that a previously unknown visitor knocked on my door, introduced himself as D.W. Parkin from the University of Bath, sat down in the armchair in my office, and started talking about his own studies, at that time none of them yet published and entirely new to me. He was sampling the dust from the Sahara carried by the Trade Winds right across the Atlantic as far as the Bahamas, where he arranged to catch it on filter papers for microscopic examination. He also told me how, by examining such material deposited on the ocean bed at different points along the way and at different times past, changes in the prevailing strength of the Trade Winds between the ice ages and the warmer, interglacial periods were revealed. At first, my anxieties about getting my own work done were only sharpened by this unplanned conversation. But soon I was mesmerized. Parkin's tale was certainly not immediately relevant to the changes that were concerning me and the inquirers whom I had to answer, but his enthusiasm and lucidity, together with his quiet, soothing voice, gained the upper hand, even in the circumstances confronting me. I quickly became convinced of the worthwhileness of his studies, which should surely tell us something about the mechanisms and effects of climatic variations on a range of time scales.

It was about this time that Sutton put me forward for a special merit promotion in the Meteorological Office without administrative duties, to free me officially to concentrate on my research. In reality, I had already enjoyed that kind of position for a number of years, but the formal recognition and status promised to be useful. He even groomed me for the interview, which loomed before me as quite an ordeal to face in mid-career after some decades of freedom from any such probing experience. He told me of another candidate, a good scientist, who had broken down when faced by the interview board and said that he did not know what, if anything, he had achieved or of what practical use his work could be. Happily, I got my promotion, but only two or three years later, with the growing interest in my subject and the lack of other people and institutions working in it, the ever-growing load of inquiries so monopolised my time that they brought my intended work programme to a standstill.

In 1965, Sutton retired and was replaced as Director-General of the Met. Office by Dr. B.J. Mason, now Sir John Mason, who did not seem to share Sutton's belief in the value of my qualifications or the kind of studies I was pursuing. My requests for at least one associated junior scientist to help deal with the burgeoning demand for information and advice on climatic variation, and take a part in the background research, now met with less sympathy. So, I was bound to be interested when, at around the same time, my old friend, Professor Gordon Manley, who had spent thirty years of his career producing the highly esteemed, and widely used, series of monthly mean temperature values typical of lowland sites around central England from about 1700 to today, suggested that I should succeed him as professor of environmental sciences at Lancaster University upon his retirement, which was due in 1967. I went to Lancaster to have a look at the possibilities but felt that the administrative and teaching responsibilities in the department would effectively halt my research, when there was still all too much to do. I also felt strongly that I could not move my family from the zestful, stimulating atmosphere of our home and the schools in Guildford and Godalming - and the delightful scenery of the wooded Surrey hills and heaths - to north-west England with its much cloudier climate. There too, there were still very visible scenes of industrial decay. So I withdrew my name from Lancaster, sadly disappointing Manley. It must have been a bitter pill for him to take, especially when the direction of research in his department changed abruptly upon his departure - though this must be a common experience for research units which bear the imprint of a strong leader.

Although I decided that the Lancaster position in a conventional university department was not for me, my visit to Lancaster led me to believe that there could be ways and means of pursuing the investigations of past climate that seemed to me likely to be fruitful, if I could find a university place where I would not be inundated with the teaching and administrative work. I therefore began to look out for opportunities in the university world. In this quest I soon had the backing again of Sir Graham Sutton, who had become head of the Natural Environment Research Council. He helped me obtain an offer of a financial grant from Shell to establish a climatic research group, if I could find a university willing to give an opportunity for it. It was at this point therefore that I began to make plans for such a development.

A chance introduction in April 1967 to Keith Clayton, then still a member of the geography department of the London School of Economics, but already designated organiser of the projected School of Environmental Sciences in the University of East Anglia, at a meeting among the North Downs in Kent to look at the curious dry valleys in the chalk - formed by the surface drainage during glacial times when the sub-strata were deeply frozen - led to the solving of my problem. Keith was, and is, a live wire, and he it was who persuaded the Nuffield Foundation to join in with another financial grant. This made it possible to launch the Climatic Research Unit in Norwich in 1971-2.

Climate was still generally taken for granted, and treated as, for practical purposes, constant, whatever its short-term vagaries might be. In the early decades of the twentieth century, up to the late nineteen-thirties, the tendencies of the climate had been towards making life easier for most people in most parts of the world. From 1896 to 1938 there had been an almost complete absence of severe winters in Europe, an almost world-wide recession of glaciers (becoming evident from as far back as 1850 but gathering pace) and of the Arctic sea ice, too, from 1920 onwards, and more reliable rainfall than before all the way from the Atlantic to central Asia, as well as in the west African and Indian monsoon regions. The seemingly inevitable consequence was that these changes passed unnoticed by the generality of mankind. But, from around 1940 onwards, quite other tendencies began to manifest themselves. The great drought in the United States Middle West in the nineteen-thirties might be taken as the earliest example. And the very severe "war winters" in Europe, three in succession between 1939-40 and 1942, with some repetition in 1945 and 1947. There were also several great Scottish snows in the nineteen-fifties, which cut transport in the north, burying trains. And there were renewed glacier advances in parts of North America. These were followed by generally colder winters, especially in Europe, in the nineteen-sixties, and failing moisture supply in Soviet central Asia. Next, twenty and more years of drought and starvation in parts of Africa, north and south of the equator, continued to mark the less favourable climatic trends. And the effects of these were all aggravated by the huge growth of the world's population.

Family Responsibilities and Changing Prospects

I was increasingly asked to give lectures and broadcast talks on climatic changes and climatic history. Events were making the case for more education and research in this field, though when I was visiting Leicester University some time around 1962 to give a lecture, and was accommodated in a house with several members of the history faculty, and was asked over sherry what I had come there for, I told them I had been asked by the geography people to give a lecture on the history of our climate and that sort of thing. This brought the response: "Oh. Is climate something that has a history?", illustrating neatly how even academics tend to keep within their own watertight compartments, and bearing witness to the need to spread knowledge and ideas more widely. The precariousness of the human situation in an already heavily populated world in the face of any change of climate - towards either warmer or colder times, wetter or drier years - is doubtless still not widely enough understood.

It was obvious in those days that historians were generally reluctant - probably many were very shy - about involving themselves in a topic which demanded some scientific knowledge and handling. This stumbling-block, however, was matched by unhelpful attitudes in the scientific community. Until much more recent years, there have been few physical scientists anywhere willing to read the literature, particularly that involving historical documentary accounts of details carefully recorded at the time, about past climate. History is not their business. Similar lack of interest has all too often characterised the attitude of physical scientists to the masses of information produced by botanists examining pollen deposits and the data turned out by geologists, glaciologists, entomologists and others. These types of literature had never been part of their regular diet. But the multiplication of types of data was causing new problems and questionable ideas of how to deal with it all in various parts of the educated community. In the mid nineteen-sixties, I resigned from the Royal Meteorological Society's library committee in protest at a decision of the Council of the Society virtually to abolish its library and display only a limited selection of the latest theoretical and interpretative texts and journals on the premises.

Inevitably, however, the theoreticians and mathematical modellers of meteorology have by now long since begun to take note of the numerical results of computer handling of the growing data on the year-layers in ice-sheets and glaciers, in tree-rings and pollen deposits, and species counts of the minute sea-living creatures present in ocean-bed deposits. These various items register year-by-year changes in the prevailing temperatures, snow and rainfall, accumulation and deposition rates, ocean currents and so forth. In such records, which automatically register climatic variations without the hand of Man - though account may have to be taken of movement of the ice or of the deposit analysed from the place at which it originally accumulated - the big changes between ice-age and interglacial conditions are obvious. Such data provide a means of checking the capacity of the theoretical mathematical models to explain the nature of changes of climate and wind-flow that observations show.

The most important result from these studies is that the timing of the great changes of climate and environment seems to agree with the changes in the sun's heat supply to the Earth, which should be expected from the well understood, astronomically caused, slow, cyclic changes (a) of tilt of the Earth's polar axis to the plane of its orbit round the sun, (b) of the elongation (ellipticity and "eccentricity") of the Earth's orbit, and thirdly (c) the slow onward march (rotation in space) of the orbit itself. This last item means that the time of year when the Earth is nearest to the sun undergoes a progressive, cyclic change. The axial-tilt cycle (a) takes about 40,000 years, the ellipticity cycle (b) nearly 100,000 years, and the precession cycle (c) 21,000 years. The astronomical measurements are now good enough to make it possible to calculate the magnitude of the changes to the seasonal heating of the Earth that should be expected at, say, thousand-year intervals at each latitude, from these variations. The climatic effects must, however, be modified and delayed by the gradual growth and decay of the ice in the polar regions. Presumably, also, some complications must be expected when, from time to time, this or that sea channel becomes blocked by ice, holding up the interchange of water with other latitudes, and at some later stage is released by break-up of the ice-jam. The verification which these studies have provided in recent decades of the astronomical basis of the broad features of the succession of ice ages and interglacials, over the era in which the large-scale geography of the Earth has been essentially as it is now, gives some confidence in predicting that the next full glacial (ice-age) climate will come about 60,000 years from now. But, unless counter-effects due to human activity supervene, the decline from the warmest post-glacial climates about four to seven or eight thousand years ago is due to reach a first, modest climax of cold, more or less glacial, climate just three to seven thousand years hence. The astronomical basis of the recurring ice ages and interglacials was strongly presented by the American participants in our first international conference in May 1973 held in the Climatic Research Unit, which I had been enabled to found, in January 1972.

Thus, the biggest swings of climate in the past may be well enough understood, and their sequels predictable. But there is as yet no comparable understanding - far less, certainty - about the causes of the shorter-term variations of climate, several of which have been great enough, and sharp enough, to have had drastic effects in history, some of them even within our own time.

CHAPTER 19. FIRST TASKS FOR CLIMATIC RESEARCH

When the Climatic Research Unit was founded, it was clear that the first and greatest need was to establish the facts of the past record of the natural climate in times before any side effects of human activities could well be important. A world-wide record was needed, particularly on the time scale of human history - a project which, surprisingly, no other body had attempted in any co-ordinated way. There was only one other similar, institution anywhere else in the world, the Center for Climatic Research set up by Reid Bryson in the University of Wisconsin at Madison in the nineteen fifties, which soon became the nucleus for an Institute of Environmental Studies. On arrival in Norwich, I was severely shocked to discover that our efforts still had not brought in enough funds to employ any staff besides myself for a contract lasting more than three years. This made us almost entirely dependant in those initial stages on whatever research on any topic might be commissioned by outside funding agencies. It soon turned out to be very difficult to attract the money needed for a programme of systematically establishing the past record. We are living in a time when the glamour of the much more expensive work of the mathematical modelling laboratories, and the tempting prospect of their theoretical predictions, are stealing the limelight. The confidence generally characterising the pronouncements from those quarters has since given way to more cautious statements in later years. It does not seem to have been widely recognised that the theoreticians' work was proceeding without adequate prior study (or any sure understanding) of the sometimes drastic swings of climate that have occurred over periods from a few years or decades to some centuries, often setting in abruptly and some of them still unexplained.

The Climatic Research Unit was becoming seriously starved of funds in its first two or three years of existence, because our applications for research grants for this or that project were more or less consistently turned down whenever we applied to any British government agency for funding. In 1974 this predicament was mentioned in an editorial note in the leading scientific journal, *Nature*, and this sparked off events which saved the situation. I learnt a few months later that the *Nature* article was passed round US government offices in Washington, DC with a hand-written memo saying "What can we do about this?". Presumably as a result, I soon received an invitation to apply to the Rockefeller Foundation in New York for funds. The Wolfson Foundation in the United Kingdom at this point also handsomely came forward with the first of several financial grants which the Climatic Research Unit has by now had from them. The later ones, received some years farther on, provided us with a purpose-built building on the University campus in Norwich. These experiences made me realise that, whatever there is to be said for the security conferred by government funding of ventures such as ours, one cannot afford to rely on just one source of support, particularly one enjoying a monopolistic position.

Two senior representatives of the Rockefeller Foundation came over from the States to Norwich to discuss with me and with the University authorities the whole question of funding, and what we hoped to do with it. The young professor, Brian Funnell, who was at that time Dean of the School of Environmental Sciences told me afterwards that he received "quite a grilling" from the visitors. My good wife prepared a nice lunch for them in our old converted farmhouse in the tiny Norfolk village of Ketteringham. It was a sunny summer's day. Soon after their arrival, we sat down a little nervously for a glass of sherry and conversation that was bound to be somewhat critical for the development of our affairs. But as soon as we had settled in our seats, the curtains which had been drawn to cut down the brightness of the sun parted and our big grey tom-cat appeared with a live rabbit in his mouth, which he briefly let go in the long room (that had probably once been the byre of the farm). Luckily, George recaptured his rabbit without bother and was shown out into the garden again. But the ice had been broken, amid laughter all round, and the conversation flowed freely from then on. We got our grant.

I was invited to a conference on climatic research that was held in June 1974 at the Rockefeller Foundation's villa - it could have appropriately been called a palazzo - beside Lake Como, at Bellagio, in northern Italy, where the warden in charge, Dr. Olsen, encouraged me to apply for a residency at some future date to help with the writing of my books. It was already becoming clear that I would have much cause to be grateful for the Rockefeller Foundation's support.

The research project which I put forward to the Rockefeller Foundation was awarded a handsome grant, but it sadly came to grief over an understandable difference of scientific judgement between me and the scientist, Dr. Tom Wigley, whom we appointed to take charge of the research. In retrospect, this difficulty might have been avoided if Dr. Wigley had been consulted at a much earlier stage on the design of the research. The scheme had been to extract the information given in the wealth of descriptive reports of the nature of individual past seasons available in a range of published collections - some of them quite famous - covering many parts of Europe and occasionally beyond, year by year over the last five hundred to a thousand years and more. My plan was that these reports should be entered on maps of the reported weather character that prevailed in the individual seasons - including the resulting harvest, floods, droughts, frosts, snows, and so on. Any relevant supplementary information published on tree-rings, thicknesses of the year-layers in ice-sheets, glacier advances and retreats, river floods etc. was to be codified and plotted on the maps also. I showed one or two sample maps for individual seasons and years of dramatic character to indicate what might be derived in the way of outlining probable wind-flow patterns to explain the reported weather.

It has seemed to me that, whereas very striking maps of some dramatic seasons could readily be produced in this way, a very long-term strategy might be needed to achieve anything like the full potential of this line of research, since the great collections of seasonal reports that were to be used were already known to contain errors. Many of the errors occurred in times before the invention of printing and were caused by copying mistakes, but others were due to muddles and mistakes about the calendar. Different countries went over to the modern (Gregorian) calendar at very different dates. The change was made in March 1582 generally in the Catholic countries and in 1700 in most Protestant countries, but not until 1752 in Britain and Ireland and as late as 1917 in Russia. The change had become necessary because the length of the day that had been used was not accurate and the effect was that a slow drift of the equinoxes and of the time of arrival of spring and the other seasons had become noticeable since the time when the older (Julian) calendar had been inaugurated by a decree of Julius Caesar in the year 46 B.C. and slightly modified about 7 B.C. by Caesar Augustus. The correction needed was finally made by introducing leap years having 366 days through the addition of a 29th day to February in the years so designated. Leap years were to occur every fourth year except in those years which begin a new century. Nevertheless, every fourth century should start with a leap year, with the exception of the starting years of every fourth millennium: thus the years 4000, 8000 etc. are not to be leap years. To make the correction needed for the "slippage" that had occurred since the Julian calendar was adopted in 46 BC, ten days had to be omitted in 1582 when Pope Gregory XIII enacted the new calendar by decree. Britain and Ireland had to omit eleven days when making the change in 1752. (This was enacted by decreeing that the day after 2nd September 1752 was to be 14th September 1752. Numbers of people, who did not understand the problem, believed they were being deprived of eleven days of their life by this decision and rioted in various places in England.) And, by the time that Russia adopted this calendar in 1917, thirteen days had to be omitted.

Other confusions were caused by the different dates that were formerly used in different countries to define the beginning of a new year. These led to mistakes about which year some weather reports really referred to. When the new year was defined as beginning in March, the whole winter would of course be allocated to the previous year, whereas the modern convention is to ascribe the winter to the year in which the January and February fall. Even more obscure, and potentially baffling, are those cases where the chronicle used reports of weather events as occurring in the xth year of the reign of some prince little known outside his own country. These difficulties have sometimes led to different dates being given for the same event in neighbouring territories and sometimes to duplicated recording of the same event in neighbouring years in neighbouring territories.

Many of these cases can, however, be resolved - although enormous care is needed and in some cases doubts remain - because Europe was formerly made up of a welter of little states, generally smaller than today's countries, which together provide more reports of past weather than any other part of the world can offer. By placing them all together on a map many unreconcilable differences between quite closely placed neighbouring reports are likely to be spotted. Some researchers, even so, may decide that the dangers of being misled are too great and that many of the reports which have come down to us must be rejected. And, in any case, the data and the maps need to be discussed and agreed by a panel of experienced people well informed about the historical background to these problems. The judgements about which reports can be accepted, and which must be rejected as misleading, are analogous to many decisions that had to be taken on daily weather maps in the past - and which are sometimes still necessary - owing to transmission errors in weather messages.

If we are ever to salvage all that can be gleaned from the collected reports of seasons in the historical past, it has seemed to me that this sifting process has to be faced. And it will require a second stage of the research with much greater funding even than we were offered by the Rockefeller Foundation, though I believe it may be a reasonable hope that the historical maps which our first efforts produce, especially for seasons of dramatic character, may generate enough interest to make the case for such an attack. The study of such maps can be expected to throw light on the long-term workings of climate (and the atmospheric and ocean circulation processes that produce it) as well as opportunities to test the impact of known solar and volcanic disturbances.

The possibilities of the mapping technique suggested had, in fact, already been tested, with encouraging results, by being applied to the summers and winters between 1680 and 1750. The first maps for the Januarys and Julys of the nineteen-twenties and thirties were drawn with only a scanty selection of data from a network of sparsely scattered points in Europe such as was available for the years 1680 to 1750. Then they were re-analysed with full information so that the errors and shortcomings could be studied. And more recently the severest part of this period has been analysed again by an international group organised by Professor Christian Pfister of Bern at a meeting held in the Danish Meteorological Institute in Copenhagen in May 1993 and attended by John Kington representing the Climatic Research Unit. (The results of this special meeting were published in a book entitled "*Climatic trends and anomalies in Europe 1675-1715*", Burkhard Frenzel, editor, Akademie der Wissenschaften und der Literatur, Mainz and European Science Foundation, Strasbourg: Gustav Fischer Verlag, Stuttgart, Jena & New York 1994. 480 pp. (See especially the chapter on "Synoptic interpretation of monthly weather maps of the late Maunder Minimum (1675-1704)" by H. Wanner, K. Frydendahl, J. Kington, C. Pfister, E. Wishman and others, pp.401-424.) Another similar experiment aimed at understanding the very dramatic weather anomalies all over Europe and the reported distress of the human populations then living between AD1300 and 1450, following hard upon the end of the most genial period of the warmer medieval climate, which was published by me in the *Beiträge zur Physik der Atmosphäre (Contributions to Atmospheric Physics*, vol. 60 No. 2, pp.131-143 Wiesbaden (Vieweg) in 1987, also produced interesting results. The most noteworthy feature indicated in the decades that were best covered was the consistent prominence of high pressure dominating the Arctic regions north of Iceland in the summers of the decades 1310-19, 1330-9, and only rather less so in the summers of the 1340s, 1420s and 1430s. During all these decades except the 1330s and 1420s the situations indicated favoured prevailing easterly winds over the northern North Sea north of about Edinburgh. In the winters of the 1420s and 1430s also a far greater frequency of easterly winds is indicated over northern Europe, including Iceland and Scotland, and the Arctic seas in this sector, than known in any later times.

    These features seem entirely consistent with the widely reported symptoms of desperately shocking breakdown of the previously ruling medieval climatic regime of which the one ominous forewarning seems to have been an increase over about a century previously in the incidence of occasional great storms. The decades here mentioned in the previous paragraph included dreadful famines across Europe between 1310 and 1319, especially in and after 1315, and extraordinary wetness of the summers in the 1340s until great heat and drought followed in 1348, when the devastating plague, the Black Death, came to Europe. Other runs of very wet years followed, with much cattle disease and many human diseases also. And in the 1430s there came an unparalleled succession of winters with long spells of severe weather. It is undoubtedly of interest that these decades fell in one of the most peculiar periods of the sun's behaviour, the late Maunder minimum of sunspot occurrences. The radically changed wind-flow pattern from the prevailing westerly winds of middle latitudes over the Atlantic and Europe, which is generally regarded as normal, seems also to offer an explanation of the evidence of over-all surprising wetness in those times over eastern England, notably Norfolk, which contains some of the driest districts nowadays and in most other times, with local courts concerned with drainage disputes. This may well be connected with the historical switch-over of agriculture in the Norfolk Breckland from grain growing in the earlier middle ages to sheep rearing. At the same time, the rainfall pattern seems to have become dislocated in other areas, for instance with the important medieval port of Sandwich on the Kent coast silting up and water-supply problems in the deep valleys in the Italian Alps south of the main ridge that caused an elaborate network of water-supply channels to be built, including one with wooden ducts designed to carry water south over the main ridge.

Despite the disappointments over the intended map series, the Rockefeller Foundation's directors were so good as to grant me resident scholar status at their Villa Serbelloni at Bellagio for some weeks in the autumn of 1978, upon my retirement from the job of directing the Climatic Research Unit. This was a richly appreciated opportunity that made possible the planning of my book on *Climate, History and the Modern World*, which was published by Methuen in 1982 and has appeared in a second edition in 1995.

In its first years, under my direction, the Climatic Research Unit quickly built up a reputation around the world for its pioneering studies of the history of climate and reconstructions of the relevant patterns of wind flow and weather. Through this, it became one of the best known departments of the then quite new University of East Anglia and even helped to make the new university (founded in 1962) itself better known in the world. Another result was that we have had opportunities to make some contributions to understanding the climatic regime in the time of cold climate in recent past centuries that has become known as the Little Ice Age.

A communication from a businessman in northern Ireland, who was something of an amateur historian with a special interest in the Spanish Armada had fascinating consequences. Pursuing that interest (which doubtless helped to keep his mind off the politics of northern Ireland), he knew what was to be found in Spanish archives in the way of logs of those ships of the Armada which escaped, and got safely back to Spain, as well as in the correspondence of some of the ships' commanders. He carefully compiled the Armada's own reports of the weather which they experienced and himself sketched proposed maps of the weather situations. He then contacted the nearest Meteorological Office staff. Together, they produced sketch maps of the proposed weather situations on over sixty days between early May and late October 1588. By a lucky chance, none of those who produced these maps - which roughly covered the area from the ocean fringe west of Scotland and Ireland, across the British Isles and surrounding seas, to near Denmark - was aware that the famous Danish scientist, Tycho Brahe, kept a daybook of wind and weather observations on the island of Hven in the Sound between Denmark and Sweden, from 1582 till 1597, including all through the year of the Armada. These observations could therefore be compared with the suggested weather maps. The maps compared reasonably with the observed weather on 72 per cent of the days, if one allowed minor position errors of fronts and weather systems by up to just 150 nautical miles. I considered this a striking success, and it made me feel justified in spending time on re-analysing their whole series of maps, with the Danish, and a few other observations added, to maintain the most rigorous continuity of interpretation of the developments. Probably the re-analysed maps give a sound explanation of the observed weather over the area mapped on eighty or more per cent of the days - an almost day-by-day account of the weather situations over this area during nearly half the year 1588. The maps were good enough to show the positions on most days of the storm/depression centres. What these maps also showed was again, as in the severe climate of the times between 1300 and 1450, a much higher frequency of easterly winds in the latitudes of Britain, the North Sea, Denmark, and the nearer part of the German plain than in the twentieth century. Also, because there is a relationship between the speeds of advance of the storm centres and prevailing wind speeds in the lower atmosphere, we can say that the maps support the impression already gained from accounts of the Armada's experiences that the winds in this region during that stormy summer were extraordinarily strong. In fact, we deduce from measurements on these maps that no less than six times in that one season, between late July and the September equinox in 1588, the winds at jet stream level over the British Isles-North Sea region exceeded the probable maximum winds at that time of the year indicated by modern wind statistics (i.e. on 1950 to 1970 figures).

Reporting of the now very surprising occurrence in the great winter of 1683-4 of wide stretches of ice off the English and French coasts of the Channel and the North Sea coasts of the Netherlands and East Anglia has in recent years been amplified by the re-printing in Dr. G.T. Meaden's *Journal of Meteorology*, vol.11, No.105 (1986), of excerpts from the diary of an observer at Lydd, Kent of how the ice, "in flakes joined so close together and where I put my staff between them I felt (more) ice underneath. Old ....observed some flakes ...... begin to come about 12 dayes (sic) before ....", was moving "as fast as I could ride foot pace" (presumably riding on horseback is referred to) seems to indicate an anomaly of the sea and ocean currents as great as that of the winds. It is now known that the Arctic sea ice about Iceland had increased greatly in those years, especially after 1675, and was affecting all the coasts of the Norwegian Sea. The complete failure of the Faeroese cod fishery, and the near-disappearance of the cod even in Shetland waters, suggests flooding of the ocean surface in that region by cold, Arctic water with prevailing temperatures around $2°$ C or less.[9] In view of these facts, it is at least possible that large amounts of the Arctic sea ice from the north may have been driven into the North Sea around that time, having passed southwards between Iceland and Norway. Indeed, this seems the likeliest explanation.

Among the efforts to attribute causes to the severe phases of the climate at that time and in the fourteenth and fifteenth centuries, there are those which more or less confidently put it all down to the known peculiar behaviour of the sun during the so-called Maunder minimum of sunspot activity that lasted from about AD 1675 to 1715 with very few signs of any solar disturbance.

Other activities of the Climatic Research Unit from its early years have been connected with conservation of natural resources, notably concerning the availability and use of wind energy.

But probably what has done most to secure the reputation of the Climatic Research Unit has been the international conferences which we have hosted and the publication of their proceedings. Difficulties had arisen over arrangements for the planned world climate conference announced for 1975 to be jointly sponsored by the World Meteorological Organisation and the International Association of Meteorology and Atmospheric Physics. The originally intending hosts at the University of Mexico could no longer do it. Luckily, we were able to save the day by providing facilities at the University of East Anglia. The participants came from all the leading countries of the world, and one Japanese professor will be remembered for getting to his feet and photographing every diagram that appeared on the screen during the six days of meetings. The proceedings were quickly published by W.M.O. as a 500-page book around the end of 1975.

---

[9] This estimate depends on the observation that at water temperature below $2°$ C the cod's kidneys fail.

First Tasks for Climatic Research

At that time, most attention was focused on the cooling of world climates that had taken place since around 1940, although the last three or four winters in Europe had been mild, a trend which was to continue until about 1978. There was also-concern about the possibility of human activities upsetting the balance of world climate.

In July 1979 the Climatic Research Unit, with the full support of the university and sponsorship from the Ford Foundation, the National Science Foundation (USA), the United Nations Environment Programme, and the World Meteorological Organization, and other bodies including the Royal Historical Society, the Prehistoric Society, the American Meteorological Society, and the Association Internationale d'Histoire Économique, arranged a week-long conference on Climate and History, our biggest international conference thus far.

My own work programme had been disturbed in the mid nineteen-sixties by a telephone call from the *Observer* newspaper's science correspondent reporting a massive Soviet scheme to "reverse the flow" of the great rivers of northern Russia and Siberia by civil engineering works on a huge scale, using nuclear explosions to blast the rocks and make new channels for the rivers Pechora, Ob and Jenesei (Yenesei), leading their waters south to replenish the increasingly arid lands of central Asia and south-eastern parts of European Russia. In those areas the inland seas, lakes and rivers (the Caspian and Aral Seas and the Amu Darya) had been drying up during the twentieth century, owing to over-use of the waters of the Volga and other rivers for irrigation as much as through declining rain and snowfall. Advantages claimed for the scheme included saving the Caspian's sturgeon and other fish-stocks and stopping the north-flowing rivers "running to waste' in the Arctic Ocean. Unfortunately, the whole scheme risked upsetting the balance of nature through damaging side-effects, which had indeed been pointed out by Russian meteorologists, notably O.A. Drozdov, in works published over many years. (I was somewhat concerned lest the publicity given to my interview in western newspapers from central Germany to Los Angeles might prove embarrassing to Drozdov, but what was noticeable was that Drozdov's work began to be published in Russia again after a gap of some years.) The diversion scheme obviously appealed to the leaders of the Soviet state because of the opportunities it offered of demonstrating what the vast scale of engineering operations that could be organised by the continent-wide dictatorship would be able to achieve. Hence, the cautions voiced by their scientists were not very welcome. This had already resulted in a long history of changes of plan, decisions that it was too risky to go ahead being followed by reversals and a rather short-lived burst of boastful propaganda. What it turned on was that the waters of those rivers maintain the low salinity of the water layer on the top of the Arctic Ocean, in which ice readily forms. If that water were not there, the deep ocean in the central Arctic could all be involved in deep convection: surface cooling would rapidly be dispersed to the depths. The dilemma is a real one. No-one can doubt the importance of conserving water for the heartlands of the world's greatest continent, but the consequences of too bold a scheme might be no less than a radical change of the whole northern hemisphere's climate. And, with a warm ice-free Arctic, (it's surface maybe 20° to 25° C warmer in winter than now) the cyclonic storm activity and the moisture supply it brings from the Atlantic and Pacific Oceans might be diverted altogether away from the continent and its drought region to the highest northern latitudes - thus making worse the very situation the works were meant to cure. In the event, it seems that sensible counsels have prevailed and just a very limited proportion of the total flow of the rivers, between 5 and 15 per cent, has been diverted, and a careful watch for adverse consequences has been kept.

Another alarm exercising the world's press had also changed the course of my work for some time in the nineteen-sixties and has caused very serious concern around the world since. This started in 1961 when the rainy season in eastern equatorial Africa proved so much more abundant than for decades previously that the level of the great Lake Victoria - whose size is comparable with the area of Ireland or the southern North Sea - rose by more than one metre within three months and around two metres by its peak height in 1964. At the same time rainfalls were generally less than in the previous thirty years at most places across Africa in latitudes between 10° and 30° N and between 10° and 25° to 30° S. Something resembling the drought features of this pattern prevailed over the next three decades after that, with a continuing rainfall deficiency that provoked a serious international emergency with great famines in Ethiopia, Eritrea and the Sudan especially, where, in the face of deaths of people and cattle in huge numbers, emergency aid was called forth on a scale never seen before.

What seems to have underlain these developments is that the two rain belts in the equatorial rains system have not migrated seasonally as far north and south of the equator in the last few decades as they did earlier, leaving rainfall deficient in the areas no longer reached by the regular seasonal rains. This change has been linked by recent investigators to sea surface temperatures in the equatorial zone, particularly of the wide Pacific Ocean, which had great influence on the world's heat economy and therefore on the global wind circulation because of its size and its particular ocean currents, with the thermal pattern which they generate. There were serious summer droughts in the nineteen-eighties and nineties in temperate latitudes over America, Europe and Asia. Increases of rain and snow downput over Arctic and sub-Arctic latitudes seem also to be part of the same global pattern and have produced increases of accumulation on the ice-sheets and some glaciers in those latitudes.

The really striking feature about the Pacific Ocean in a normal year is the relatively low temperatures of the surface water along the equator, commonly about 24° C to 27° in the eastern Pacific to as far as the date line (180° W or E), but occasionally only about 21° and even as low as 18° to 20° near the South American coast (unlike the general ocean surface values near the equator of 28° to 30° in other sectors). The cool equator in the Pacific comes from the cold Humboldt current moving north along the Chilean coast from the Antarctic. But this cold water supply weakens every few years and allows warmer water from the north side of the equator to bring temperatures along the equator in the Pacific for a few months to a year up to the level normal in other sectors. These gross variations affect the development of the winds and weather in many other parts of the world. The arrival of the warm water along the coast of Peru (up to 4° to 6° C above the normal for those places) is inclined to set in about Christmas: it has become known therefore among the Catholic peasantry as El Niño (the Baby). This name is now very widely applied by meteorologists and oceanographers around the world to the change in the broad-scale pattern of ocean currents and all its effects. The effects are liable to include gross changes from the usual drought of the Peru coast to heavy rainfall there and in Ecuador and the Pacific islands near the same low latitude range. In another realm, the change for the time being of the water mass along that coast had brought complete collapse of the usually very profitable Peruvian anchovy fishery. But if it is these warm water years that bring the occasional great rains in that region, those same years also tend to be associated with colder than usual water in the equatorial Atlantic and droughts across Africa that have caused serious emergencies for the populations in that continent, not only in the Sahel, Ethiopia, Eritrea and Somalia, but also south of the equator in Mozambique and its neighbouring territories. These emergencies have provoked urgent international appeals for aid in the nineteen-eighties and nineties on a scale never before organised. There seems also to be an association between the El Niño years and an enhanced tendency for high pressure and blocking anticyclones to develop over sub-Arctic latitudes and some regions in similar southern hemisphere latitudes.

There were great El Niños in 1982-3, 1972 and, perhaps, 1957 as well as, according to report, in 1877. Latterly, a much more persistent El Niño has occurred, almost continuous from 1990 onwards, which might be looked upon as an aspect of general warming of the lower latitudes of the Earth.

All these interwoven climatic developments are of great concern to humanity, especially in low latitudes, where populations have long been growing fast and food and water supply problems are much aggravated thereby. These alarms keep the demand for climatic research and advice on the boil.

In the nineteen-sixties, the German Navy reported concern nearer home over an observed trend, from about 1953 on, towards increasing roughness of the North Sea, which was affecting its operating areas. Since that time increasing windiness has been reported on the west Swedish coast and in many parts of the world farther away. This tendency is shown alike by the average wind speeds and the frequency of storms. Professor Hermann Flohn of Bonn has argued that this trend could have been expected from the observed warming in recent decades of the tropical oceans and to some extent of other oceans elsewhere, particularly in southern latitudes. The warming of the sea surfaces in those latitudes has averaged about 0.3° C over the thirty years after the nineteen-sixties. One can be sure that this warming must have led to increased evaporation from the ocean surface and therewith an increase of moisture content generally in the atmosphere. Condensation of that moisture to form clouds must release latent heat and therewith extra energy for the winds. Flohn has estimated on this basis an over-all intensification of the Earth's wind circulation in recent years by probably 10 to 12%. Sir Solly Zuckerman, who was government Chief Scientist for many years in the nineteen-sixties and seventies, and later became Lord Zuckerman, was quick to fasten on this increasing windiness and its implications, which led him to promote the building of the Thames Barrier to protect the city of London, especially in case of any further rise of sea level. He also became a firm supporter of the Climatic Research Unit.

An investigation, which I was enabled to carry out through collaboration with my friend Knud Frydendahl of the Danish Meteorological Institute, using that authority's splendid collection of computerised ships' observations, also shows this recent trend in its setting against a long back history. Danish and Norwegian ships have the advantage for this study that they frequent the stormiest regions of the Atlantic Ocean north of 50° N more than the ships of other nations. Moreover, the collection extends back to the seventeenth century. Thereby, it also reveals other periods marked by very great storms in earlier times for which the meteorological situations causing the storms could be reconstructed from the network of observations available. This study reveals that another period of enhanced storminess occurred in the time of generally colder climate, now generally known as the Little Ice Age (witness the lessons of the Spanish Armada study!).

It might be argued that this storminess of the time of cold climate could be no more than a feature of the latitude zone of the seas off northern and western Europe and around the British isles (which were the main region of study), on account of sharpened temperature contrasts around the Arctic fringe at that time. But we cannot be quite sure that it was such a localized feature, and we cannot therefore be quite sure of the cause.

We can glimpse a longer history of the storminess of the North Atlantic from the records of serious sea floods on the coasts of Europe over the last two thousand years. Of course, one cannot suppose that the record is complete, even though we concentrate just on disastrous floods. Nevertheless, it shows a remarkable thing: despite the probable losses of documentary records through the ages, the numbers of reported severe sea floods on the coasts of the North Sea and the Channel in the eleventh and, especially, during the thirteenth century AD greatly exceed those in any century since. In the case of the thirteenth century floods, the numbers of serious cases are almost three times as great as in any century since. This coincides with what appears to be the culminating stage of the medieval period of warmer climate, which seems to have been most pronounced in the northern hemisphere. Presumably, world sea level was at least somewhat raised thereby (through the melting of glaciers).But probably even more relevant is that that time was the final stage of the warmer climate: from about AD 1190 onward there were signs - e.g. increasing appearances of the Arctic sea ice - of cooling setting in the highest latitudes and therefore of increasing thermal contrasts (gradient) between north and south, which would supply the energy for a stronger wind circulation.

The records indicate continual variations of storminess in the past but no simple pattern for any one area or region. Over all, if we consider the world as a whole, we must suppose that the greatest energy of the winds will be manifested when the most water vapour is in the atmosphere, available to release the greatest store of energy as the latent heat of condensation when and where clouds are formed. Clouds have to be expected to form most in regions where thermal contrasts exist. But in the records for individual regions - as, for example, the North Sea, British Isles and northernmost Atlantic, where we investigated the history from the relatively abundant records which exist in this part of the world - there have been climaxes of storminess in and around the warmest and the coldest periods in the past and particularly near the culminating stages of either development.

There was another activity in these years that involved me in a certain amount of travelling, membership of and in one or two cases the chairmanship of international working groups, mostly appointed by the World Meteorological Organisation. Such groups are constituted to look after the essential business end of international co-operation and their work is not necessarily of great scientific interest. There was just one case of wider interest that I was involved with. This was the Polar Meteorology Group whose business was to look after the fulfilment of internationally agreed scientific effort in the organising and collecting of meteorological observations from places in the Far South, in the Antarctic and sub-Antarctic zone. Britain had responsibilities for the provision of observations from posts on and near the coasts of Antarctica in the so-named Falkland Islands Dependencies region. But my efforts to remind the authorities of this in the late nineteen-seventies and round about 1980-81 met with a sort of disbelieving response, implying that, unless I was deliberately out to make trouble, I could not really be serious about the matter. All that changed overnight when the Argentine forces invaded the Falkland Islands! One wondered whether our political leaders even knew where the territories named were until the invasion was imminent. Their lack of interest in the whole area had been so marked and was clearly a matter of policy.

One other incident from this realm of my activities sticks in mind for quite different reasons. It was again the Polar Meteorology Working Group that I was attending, though the meeting place was Paris in September in the nineteen-seventies and the weather was warm. There were seven members of the group from six or seven different nations, including Russia and the United States, and our work had made us rather late in going out for an evening meal. So we sallied forth to find a pavement restaurant on a boulevard not far from where we were working. We sat around a rather wide round table and, after discussion, ordered our meal. The chairman of the working group, Kaare Langlo, a Norwegian with much experience of working at W.M.O. headquarters in Geneva and of membership of other working groups meeting in Paris and elsewhere, asked to see the wine list and picked from it one particular French red wine. But the waiter tried so hard to persuade him to choose a different French red that, being faced by the evident readiness of the rest of us to acquiesce, Kaare agreed. But, when the waiter returned and poured a little out of the bottle for him to try, he pronounced it "not good" and insisted on having the wine which he had originally asked for. So a bottle or two was brought. But, in opening the first one, the waiter, now appearing nervous, spilt at least a glassful of it down the front of Kaare's smart blue blazer. In the furious row which followed, he insisted on having his blazer cleaned at the restaurant's expense. How well he succeeded in the outcome the rest of us never heard.

First Tasks for Climatic Research

One can only welcome, indeed as long overdue, the increasing concern world-wide nowadays about care and conservation of the environment and the using up of limited resources. This is a matter of leaving a world in which our descendants may have the chance to enjoy happy and secure lives. This precept was glimpsed, and laid upon our forebears, long ago in the books of Moses and the prophets in the old testament of the Bible. The situation facing mankind today is made much more urgent by the galloping increase of the human population and the exhaustion already of some resources of land, minerals and food. The outlook for the climate and for all aspects of life - notably food production and transport - has seemed to many people, including some scientists, a great deal simpler and more obvious than it really is. We should be warned by the record of alternations of view between acceptance and doubt about global warming. It was no less a figure than the late Reggie Sutcliffe - Professor Sutcliffe, former director of research in the Meteorological Office - who remarked at a meteorological dinner around 1970 that it seemed "almost impious to suggest that we could create mathematical models" of the atmosphere and climate that would express, and give the right weight to, all the influences on the system that matter.

What is certain is that we should "walk humbly over the Earth". We must proceed with caution and a lively awareness of the possibilities of precipitating disaster, now that the magnitude of our impacts on the planet has vastly increased. The most obvious - though not obvious to everybody - of those possibilities are related to the accumulation of essentially everlastingly active nuclear wastes.

In 1896 the Swedish scientist, Sv. Arrhenius, professor of physics first at Uppsala and later in Stockholm, published his suggestion that increasing the carbon dioxide in the atmosphere, as was already happening relentlessly, should be expected to warm world climates because of its absorption - i.e. capture - of long-wave radiation that continually goes out from the Earth and so create a sort of "greenhouse effect". And in 1938 in England G.S. Callendar seemed to show in a paper in the Royal Meteorological Society's journal that the observed warming of surface temperatures over the Earth by about half a degree Celsius from around 1890 to the nineteen-thirties should be about right to be attributable to the radiation trapped in the atmosphere in this way. But there are some difficult points. Water vapour, which is abundant in the atmosphere except over the coldest regions of the Earth and in the stratosphere, also absorbs radiation and on almost all the same wave-lengths that the carbon dioxide absorbs. Difficulties, too, beset attempts to show how variations in the amounts of carbon dioxide in the atmosphere in the past fit the theory that warm periods in world climate can be attributed to a greater CO_2 content and cold periods to a lower CO_2 amount. The CO_2 content at various past times is presumably indicated by the gas trapped in bubbles in ice-sheets and glaciers. This does show less CO_2 in glacial times, and during warmer interglacial periods the CO_2 amounts were greater. But, since carbon dioxide is more soluble in water - in the oceans for example - when temperatures are lower, the smaller amounts of CO_2 in the bubbles in the ice sheets in ice age times could be just a result of the colder climates then prevailing. And, even within our own times, the suggestion that the increasing carbon dioxide in the atmosphere should be presumed to be the cause of the warming does not fit at all well with the sequence of observed values. The great period of warming, at least in the northern hemisphere, was during the first forty years of the twentieth century (especially the first and fourth decades), but in the nineteen fifties and sixties when the CO_2 was increasing more rapidly than ever before the prevailing temperatures were falling. Callendar himself was worried by this discrepancy and contacted both me and Professor Gordon Manley about it. There seem, in fact, to have been a number of shorter runs of sometimes up to fifty years with either rising or falling temperatures, often setting in suddenly, and with no clear correspondence to changes in the atmospheric CO_2 content.

We also see that account must be taken of psychological reactions - even in the influential research community - to the variations towards greater or less warmth as and when they occur. In the 1880s and '90s, as a recent American meteorological investigator was the first to be able to show, world temperatures were lower than they had been since around 1850. That was just when Arrhenius came out with his suggestion that the man-made increase of carbon dioxide should be warming the Earth. And at that time the suggestion made little impact. When Callendar promoted the same idea forty years later, however, it was in a warmer world, though very soon the bitter war winters came and implanted themselves in folk's memories. And when G.N. Plass again put forward the CO_2 warming theory in papers published in 1954 and 1956, world climate was once more entering a colder phase, particularly in the northern hemisphere. Interest in the theory soon waned. It only revived after a run of up to eight mild winters in a row affected much of Europe and parts of North America in the 1970s and '80s. There then came a tremendous preponderance of publications on global warming, dominating the research literature, although over-all temperature averages in some regions, particularly in the Arctic, were still moving downward. So, in spite of the sharp turn towards warming after 1987-8, and the undeniably very warm years 1989-91 and 1995, one must feel uneasy about the confidence with which global warming has been publicised as the verdict of science in official pronouncements from many quarters.

The erratic course of the changes experienced through the twentieth century surely suggests that there are processes at work which are still not adequately understood and possibly even some influences that have not yet been identified. We can at least glimpse one important mechanism at work in some of these surprising turns back to a colder climate trend in the sudden increases of the volume of cold Arctic water moving south in the East Greenland Sea and pushing the boundary of the warmer, saline water of Gulf Stream - North Atlantic Drift water south, as occurred in the nineteen-sixties. The case examined by Professor L. Mysak of McGill University, Montreal was attributable to a great increase of run-off from the Canadian rivers. The effects of it could be traced over many years. The controls of this behaviour need to be much more fully understood. The flow of the Arctic water current is in some years increased by a factor of ten, and a similar or possibly even much greater outflow of Arctic water, was a major feature of the cold climate of the Little Ice Age period in the sixteenth and seventeenth centuries.

It may be important to notice that this southward switch of the warm water boundary in the north-east Atlantic takes place in the very same longitudes as the much greater southward shift that characterized the main (Quaternary) ice ages.

Around 1960, Professor M.T. Budyko of the USSR Academy of Sciences, and his colleagues in the Hydrological Institute in Leningrad and the Voeikov Main Geophysical Observatory there, were already greatly concerned with the changes that climate undergoes and the threat of increasing influence from human activities. They placed strong emphasis on inadvertent modification - meaning, in general, warming - of world climate through the increase of carbon dioxide in the atmosphere and from other gases released by human activities which add to the "greenhouse effect" by similarly absorbing, and thereby checking, the outgoing radiation from the Earth. Budyko was reported to have suggested at a conference in Moscow in 1960 that it might soon become necessary for mankind to intervene in some ways to counter the development, possibly by using aircraft to spread veils of dust and smoke in the stratosphere, to reduce the strength of the incoming sun's beam.

So, after a long period in which the question of climatic variation was generally ignored, the subject was becoming a live one again and many international conferences were held from about 1960 onward to discuss it. Between the eighteenth and the mid-nineteenth centuries a very different attitude had prevailed, owing to advances of the glaciers in many parts of the world and the occurrence of various kinds of extreme seasons. It was indeed the anxieties in that time, particularly among those whose land was being overrun by ice, reduced to swamp, or continually threatened by landslides or rock-falls, and elsewhere by sea floods or advances of coastal sand dunes, that had led to the first officially organized networks of posts for regular meteorological observations being set up. Now, with no less than three world conferences on the subject which I was encouraged to attend in the one year 1961, the matter was returning to "centre of the stage". The first of these conferences was in late January of that year, under the auspices of the New York Academy of Sciences, jointly with the American Meteorological Society, chaired by Professor Rhodes W. Fairbridge of Columbia University, attended by six to eight or more of the world's best known names in meteorology and climatology as well as leading specialists in Earth history - geologists, botanists, glaciologists, marine biologists, and oceanographers - though there were also one or two self-revealing cranks. There were archaeologists too, who had specialised in the American West, in the Mediterranean and in Africa, as well as the Arctic shores of North America and the Soviet Union. It was at this New York conference that I first met the American Weather Bureau's charming young specialist on climatic change, Dr. J. Murray Mitchell jr. Murray had a brilliant mind and had achieved at a young age a remarkable mastery not only of dynamical meteorology but of all the statistical routines that made possible the handling and analysis of the vast quantities of observations involved in world climate and its changes. He was, with all his family, a fine upstanding member of the Episcopalian (Anglican) church in the USA, one of the best type of American gentlemen in every sense of that word, a real conservative, kindly, loyal and with the faultless manners of the traditional WASP, the White Anglo-Saxon Protestant type of American citizen. We became very good friends until his sad death thirty years later from a lymphatic cancer. But I was very shocked when he told me a year or so after our first meeting that he would be taking time out from his work to go in for war studies.

Other things provided a bizarre backcloth to that conference. It took place in a week of very bitter late January weather, including a great snowfall, which brought the powerful snow-blowing machines out to clear the streets, and as they worked their way along the famous avenues and boulevards, they buried all the vehicles parked along the sides deeply in the snow. The frost was severe, the temperatures advertized by the publicly illuminated roof-top thermometers ranged only between -8° and -15° C (18° to 5° F) all that week, and there was a strong wind. But the mostly middle-aged or older European scientists, accommodated for economic reasons all together in a huge dormitory in the basement of a university building, were severely overheated. The room temperature was kept at +28° C (82°F) except when one or other of the visitors surreptitiously opened a window (but these were always soon closed again). I could only stand a limited period in that room before I felt obliged to go out for a "breather". And of course one was very soon driven back indoors. By the end of the week, I for one developed symptoms of 'flu' and a temperature. But happily two hours on the Saturday afternoon spent in the constant, more moderate warmth of the Metropolitan Museum of Art set me on the road to recovery.

There was also a rather well known "character" from England attending the conference, Dr. D. Justin Schove, the headmaster of a "private preparatory school" in Kent, which he had inherited and which evidently ran itself without much attention from him, although it probably contributed usefully to his income. He was a great enthusiast in his interest in anything to do with climatic changes, who attended all meetings within range of London, and often farther afield, on the topic. Occasionally, he produced a useful piece of work himself. His interest in the sun and solar variations led him to publish a long series of yearly sunspot numbers and an even longer history of the characteristics of each sunspot cycle, based on Chinese records collected by Dr. Joseph Needham from about 650BC to the present. At this New York conference, Schove was in a state of great excitement throughout, standing up to make some comment after nearly every paper, and giving three or four press conferences about his own work, which he arranged himself by advance publicity. Some people present, notably newspaper men, clearly came specially to meet the great D.J. Schove. One undoubtedly useful service, which he did provide, was a routine by-product of his extensive correspondence: this was his keenness to introduce highly respected workers to each other. He derived evident satisfaction from this conference for the chance it gave him to meet the scientists with the well-known names and tell them of each other's work. He was a very kind, likeable fellow, a fearless dabbler in a wide range of studies, which he dealt with at great speed but not always with enough critical consideration. Schove attached himself to me at times throughout the week, and I found myself finally destined to travel back to England with him when it was all over, as we were booked on the same plane. I had to realise that I had better look after him a bit and make sure that he caught his connexions: his fluster and his continual state of excitement made it all too likely that he would go astray somewhere. However, the worst that befell was when his main travelling bag flew open and scattered its contents all over the floor of the subway train, right in the way of the exit, shortly before we had to change trains on our way to the airport. As it was a Sunday afternoon, there were not too many people about: so we could gather up his belongings without too much difficulty. But his luggage was made bulky by the sheaf of fat Sunday editions of the New York papers that he had bought and hurriedly scanned to see what references to his own comments and contributions to the conference he could find in them - a feverish activity that had contributed to the mishap.

In August-September of that year, I attended a big conference on this subject in Poland, which was then still behind the Iron Curtain and quite heavily under Russian surveillance. As a British civil servant, I had to go through a somewhat sinister, but no doubt valuable, briefing at the Foreign Office in London before going. One was certainly given a useful impression of ways in which the communist authorities might try to compromise one. So I duly went off to Warsaw for the 4-yearly meeting of the International Association for Quaternary Research (INQUA). Like other bodies of its kind, INQUA meets in a different country each time at the invitation of the host country, and is usually attended by some hundreds of scientists from many different countries. The Met. Office had generously sponsored me for a stay of up to three weeks, to give time to go on one or other of the field excursions announced. The reason for this was that it was proving impossible to discover beforehand which dates could be taken up by the excursions and which were chosen for the meetings in Warsaw. I put my name down for accommodation in the cheapest of the hotels recommended in the pre-Conference literature. And so it turned out that, apart from one Irish botanist whom I already knew slightly, I was the only representative of any western country staying there. It turned out that this modest choice of hotel had unforeseen advantages besides. When I met my American friend, Murray Mitchell, later that evening, he and another American who was sharing the room in the best hotel were already telling a tale of a strange experience they had had soon after arrival, just the sort of attempt to compromise one that I had been warned about at the Foreign Office in London. They had not been many minutes in the room, and were in the midst of changing their clothes, when on answering a knock on the door they were met by an attractive girl, only partly dressed, who marched in on some flimsy excuse and asked them about some probably made up problem. They had some difficulty in persuading her to leave. And they were subjected to one more attempt to compromise them later, and found evidence that their room had been "bugged". Things were simpler in the Russian dominated quarters that I had found.

On another evening when some of us came back to our hotel late for the evening meal after a long official excursion, I was sitting at supper at a sparsely occupied table near two other scientists, who were talking German, and I could not help overhearing their conversation, which gave a nice example of relations of a very normal kind behind the Iron Curtain. The younger man was German, from Berlin, and he was taking the opportunity offered by finding himself dining virtually alone with the older man, who was evidently Russian, visiting Warsaw for the conference, and who seemed to be a highly respected professor in the same line as himself. The younger man, who was clearly ambitious, had lately been appointed to a laboratory of his own and was very proud of it. He was trying hard to persuade the Russian to come to Berlin to see what he was doing. But the professor explained that "these things are not so easy yet, you know". And, although they had a long discussion and several different arguments were tried to persuade him, the Russian professor had to tell the keen young man gently, but firmly, that he really could not come.

The following week was mostly spent on one of the main excursions, in the south of Poland, in Silesia, an area that had long been part of Germany. I was sad to see the neat village houses still badly scarred by war damage over sixteen years after the fighting. The Poles had had no choice but to take this land which was offered them by the Russians in compensation for historic Polish lands farther east, that had been taken into Russia and the Ukraine at the end of the war. We stayed a few nights in Wroclaw (formerly Breslau), where daily meteorological observations were taken from about 1700. And we had a good outing on a clear day on the tops of the shapely mountains previously known as the Erzgebirge and now the Krkonose, along the border between Poland and Czechoslovakia. We climbed to the top of Scheekoppe (Snezka) a 5,000 foot (over 1,500 metres) peak on the ridge. In spite of the fact that it was mid week, there were great numbers of people on foot, in big organized parties, on every stretch of the broad path that led from summit to summit. The date was just before the end of August. We were plainly witnessing organised mass holidays for the working population. The principle was the same as the Nazis' "Strength through Joy" (Kraft durch Freude) movement, not that that constituted any argument against such a healthy pursuit. And it was noticeable that the walkers had been persuaded to leave no litter.

Among us foreign visitors to their conference there was continual amazement at the seemingly imperturbable Poles making caustic remarks about their Russian conquerors within the hearing of many of them who might be present. The most brazen, and surely the riskiest, statement that one of them made to me was that it was Poland's terrible misfortune to be placed between two such awful neighbours as the Germans and the Russians, but that of the two the Russians were the worst!

During this Polish visit we came to a place in Silesia that had formerly been German, where there was a very obviously Norwegian ancient wooden stave church. Like another grand old stave church that we already knew in the Harz mountains, this one had probably been bought by Kaiser Wilhelm II before the first world war, transported here, and reassembled. I went in and found inside just a couple of Norwegians from our conference viewing the church. I entered into their conversation which interested me, and they were surprised by my knowledge of their language. So began a long friendship with Professor Just Gjessing of Oslo which led to contacts with others working on the development of climate and landscape in Norway after the last ice age, and a few years later I gave a lecture in his geography department in Oslo University.

My own contribution to these international congresses in 1961 was primarily to demonstrate how the prevailing patterns of the world's wind circulation could be reconstructed from the actual meteorological observations made at times back to around 1700 to 1750 in Europe, and around the North Atlantic Ocean, and what related exercises could be carried out with more indirect evidence such as the frequency, decade by decade, of evidence of prevailing warm, dry, cold, and wet seasons in different longitudes in Europe. These exercises and reconstructions depended on modern understanding of the most basic feature of the global wind circulation, the great circumpolar vortex (whirl) of upper westerly winds that dominates the scene between heights of about two and twelve to fifteen kilometres above the Earth's surface. The system owes its origin to the unequal heating of the surface and overlying atmosphere in different latitudes. But it also includes imbalances that generate the travelling weather systems that are steered along by the wind flow in which they are embedded. These individual travelling weather systems can, and do, however, distort the main pattern of flow itself when they transport enough heat to do so. The pattern of prevailing weather, repeatedly disturbed or serene, windy and stormy or relatively calm, wet or dry, is obviously related to the positions of the main flow of the upper winds and its distortions. And, because the basic features of the flow of the winds in this great system are governed by the physical laws of dynamics and thermodynamics, one can make deductions about related changes when any change occurs. Thus, changes in the west-to-east spacing of prevailing wetness or dryness in Europe were understood as changes of prevailing wave-length between the ridges and troughs in the flow of the upper westerly winds around the hemisphere down-wind from the fixed disturbance induced by the Rocky Mountains barrier. These changes of wave length should be explained either by a change of latitude of the main flow which would be signalled by a change in the latitude of the main thermal gradient and zone of weather disturbances, or by evidence of generally stronger or weaker main flow. There did, for instance, seem to be evidence that the zone of storm activity shifted to the north in the time of greater warmth in the high middle ages and that the storm activity increased in the thirteenth century, towards the end of that time. The activity was seemingly displaced south during the coldest climate period between the late sixteenth and late seventeenth centuries.

What I gained from the Polish INQUA conference was particularly the chance to observe at first hand how geologists and other field scientists go about examining and interpreting their evidence of past climates and the uncertainties which they must try to resolve. Here they were observing the surface deposits of materials produced and moved about by geologically recent climates, depending particularly on the action of frost and thaw, glaciers and melt water, floods and droughts. Their actions and discussions on site were all the more valuable since some world-renowned geologists, such as Richard Foster Flint and Fred (F.W.) Shotton, were taking part. Resolution of the uncertainties often depended upon identification of evidence of movement or modification of the materials after deposition.

The final congress on changes of climate that I was enabled by the Met. Office to attend in 1961 was held in Rome in October, organised by the United Nations Educational, Scientific and Cultural Organization (UNESCO) jointly with the World Meteorological Organization (WMO), and was particularly concerned with the arid zone. The regions reviewed included Kurdistan, Iran and Iraq, and Israel, the Ukraine, northern and southern Africa. And attention was given to the work of archaeologists from Soviet Asia, besides tree-ring workers', and oceanographers' and meteorologists', work ranging from low latitudes to the Arctic. In the final session, the American meteorologist, Jerome Namias, commented that "these meetings will have been specially valuable for the theoreticians in bringing them up against the facts of climatic change, so tending to save them from continually floundering about in their ignorance". But this is likely to prove an elusive hope because of the very different types of mind of research workers in these different realms. A high proportion of the theoreticians will presumably always prefer the tidy, beautiful patterns of theory to the complexities of the real world. And they may always prefer to dismiss various occasional and supposedly marginal influences - solar?, gravitational?, tidal force?, fortuitous ice blockages?, and so on - which might spoil the regularity or the symmetry of their patterns. And the possibility remains of occasional disturbances even of the long-period astronomical rhythms of ice ages and the warm interglacial periods by sudden outflows of very great ice masses from the Antarctic continent.

In June 1962 there was a useful meeting held 8,000 feet up amid the high Rocky Mountains at Aspen, Colorado to examine the characteristics of two extreme phases of climate development in historical times, the warm period which culminated in different parts of the northern hemisphere around the eleventh century - or perhaps better described as the tenth to thirteenth centuries AD - and what is commonly called the Little Ice Age around the sixteenth, or better the fifteenth to seventeenth centuries. To get to Aspen, we were bussed in bright sunshine from the airport at Denver on a fine road up over the Continental Divide at about 12,000 feet above sea level. On the flight across the Atlantic to North America on 15th June 1962 we had a splendid, clear view of the patterns of flow of the East Greenland ice drifting - south from the Arctic Ocean, "hugging" the Greenland coast, and then being caught repeatedly into large-scale, beautifully formed, anticlockwise eddies, presumably through encountering the powerful drag of the warm Irminger Current - a branch of the Gulf Stream-North Atlantic Drift - heading north-east and north into the eastern half of the broad channel between Iceland and Greenland. The view continued clear as far as the whole eastern slope of the Greenland ice cap and the nunataks (rock outcrops and bastions) that stick right through the ice. For my return journey I found that, at very small extra cost, I could fly on west to the Pacific coast at Portland, Oregon to visit a widowed aunt and thence to Seattle. After an hour's stop there, I headed for home over the Canadian Rockies. That Pan American Airways flight took me diagonally north-eastwards over Hudson's Bay, which on 26th June was still over nine-tenths covered with ice. We alighted on Baffin Island for an hour or so that afternoon, in bright windy weather, with the temperature just above the freezing point, and then crossed southern Greenland and thence south-east towards the British Isles. We never came within sight of Iceland this time, but there was an unexpected bonus as we passed directly over St. Kilda, seeing the whole group of small islands in clear, early morning sunshine, and reaching London Airport an hour and a half later.

This was manifestly a cold summer, as our family was able to observe when we drove over the Norwegian mountains by the road from Bergen and Hardanger fjord in the west to a night-stop on the heights, about 1,000 metres up, at Haukelisaeter, where the landscape on 9-10th August was still more than half covered by snow. We drove on and were soon with our friends, the Ramlets, at Fevik on the south coast, where it was hard this year to find water temperatures much above 15° for bathing instead of the usual 18° to 22°.

In March 1964 I went to a conference of the German Geologische Vereinigung in Cologne, where I showed the pattern of prevailing sea surface temperatures in the North Atlantic in the 1840s, in one of the later phases of the Little Ice Age, based on observations collected by the

First Tasks for Climatic Research

American Admiral Maury. The remarkable feature of this, like the pattern derived from the observations made between 1780 and 1820 for the British Admiralty and collected by Rennell, is a belt of water temperatures up to 2°, and locally even 3° C, higher than in recent decades across the ocean between latitudes 25° and 42° to 45° N.

The Royal Meteorological Society itself organized an international symposium in London in 1966 on world climates through postglacial times from 8000 BC to the time of Christ. Evidence was considered from the glaciers all over the world and from sea level changes to what is known of the history of the vegetation and changes in the marine biological species represented in the deposits on the ocean bed. These approaches seem to point to a succession of climate stages, which could be defined by the positions of the limits of glaciers, ice-sheets and permafrost, and of the chief forest types, steppes and deserts. Astronomical data were used to define the changes of incoming radiation from the sun in a critical sub-polar latitude, and radiocarbon studies indicated some probable changes of solar output, whilst the marine biology indicated an outline history of changes of prevailing ocean surface temperatures. Further adjustments to the maps were needed to show changes in the position of the coastlines as glaciers melted and the sea level rose. The results indicated four important phases of postglacial time each with its characteristic positions of the major geographical boundaries. Using radiocarbon dates for the evidence, these broad divisions were:

(i) around 6000 to 7000BC, in the early post-glacial, a so-called "Boreal climatic period",

(ii) around 4500-4000BC, the so-called "Atlantic climatic period",

(iii) between about 2500 and 2000BC, the "Sub-Boreal climatic period", followed by

(iv) a marked cooling phase so far as Europe and its coasts were concerned, known as the "Sub-Atlantic climatic period".

The Atlantic period seems on botanical evidence to have marked the culmination of a very oceanic, mild and humid climate, with westerly winds very predominant. The preceding and immediately following periods seem to have had much more continental climate characteristics in Britain and Europe. The evidence from "Sub-Atlantic" times suggests a return to more oceanic, but colder, climate, presumably with many west and north-west winds.

First Tasks for Climatic Research

My part in this was to produce an account of the underlying geography at each stage and to collaborate with a Met. Office colleague, A. Woodroffe, who was deeply involved in mathematical modelling of the general wind circulation and numerical forecasting. We were of course working with the prevailing temperatures indicated to derive a picture of the average wind circulation that would correspond to equilibrium with the temperature pattern. A few years later, Woodroffe and I repeated the same procedures to derive maps for several of the main stages of the last ice age:

(i) around the time of maximum glaciation, some 20,000 years ago,

(ii) during the climax of an early warm period in the deglaciation, the so-called Allerod epoch, around 12,000 years ago, and

(iii) the last, severe cold stage in Europe, around 10,000 to 11,000 years ago.

The maps of all these stages have been printed in my book *Climate: present, past and future*, volume 2, published by Methuen in 1977, and in the journal *Quaternary Research*.

The weather provided a stimulating background to the years that we lived in Guildford. We enjoyed a succession of skating Christmases in 1961 to 1964, when the city flooded the riverside meadows, enabling a happy crowd of skaters to enjoy themselves safely on ice with no water underneath. And there were opportunities for local skiing both on the open grassland of Pewley Downs and the lovely paths through the pinewoods on the higher ground about Leith Hill and Holmbury Hill. In the great winter of 1962-3 the snowfall on 26-28th December produced a level depth of 40 cm of snow outside our house and snow clearance became impossible. We reckoned that during that one winter we and our family had several hundred pounds worth of winter sports without ever leaving home. The costs were kept low by a "white elephant shop" in Godalming, where we could sell the children's skates and other gear when they grew too big for them and buy second-hand replacements in whatever bigger sizes were required. We built an igloo on the road outside our house on the 29th, and the roof did not fall in until March 12th. There were 65 days of snow cover around where we lived in the upper part of Guildford in that remarkable winter. As we lived near the top of a one-in-five hill, I took to fitting deep-tread all-weather tyres on the driving wheels of our car each winter. The advantage was never more obvious than on a Friday (the 13th) in February 1970, when I had to drive to Cambridge for one of the botany department's Quaternary Discussion meetings. I took two neighbours from Guildford, who had their own reasons for visiting Cambridge that day, and arranged for a return in the late evening. The day was cold but fine and sunny after some frost. But then between 5 and 6 p.m. a snow-belt, bringing colder air from the Arctic seas, swept south over this part of England, with squally winds. Within an hour the whole region was under several inches of snow, which was drifting wildly in the wind, a true blizzard. It came just at the evening rush hour, and soon the roads and roadsides were littered with abandoned vehicles, each one appearing as a rounded, white obstacle. By 10.30 p.m. the time appointed when we collected our passengers for the drive home to Guildford, the weather was clear and starry, with less wind and hard frost, and there was hardly any traffic. We counted over 150 vehicles left on, or beside, the road on our way south. Our deep tread tyres made the going easy as long as we avoided the deeper snow. With some trepidation, we overtook a police patrol somewhere in East London, but we were not exceeding the speed limit. The shortest route took us through central London. The wind was still blowing the snow up from the deserted streets, so that wisps of it were even whirling past the dome of St. Paul's Cathedral. We reached home safely between 1 and 2 a.m.

    In the last months of 1969, our elder daughter, Catherine, for the second year of her languages course at the University of East Anglia, was studying German in the University of Hanover before going on to the next semester in Göttingen. We therefore decided to take the family over to join her for a Christmas holiday in the Harz mountains, at the high-level village of Hahnenklee, which we had got to know in 1951-2, when we ourselves were living in Germany. We were lucky with the weather. The snow came to northern Germany early that year, about the beginning of December. By the date of our arrival there was already a ten-inch (25 cm) depth of it, making a lovely scene in the pinewoods and in the village, where some of the houses were old and picturesque, as was the great wooden stave church from Norway, that had been re-erected there. For skiing, there was an excellent sloping practice meadow, below the woods at the edge of the village, and a fine, broad run, nearly 2 km long, through the woods down from the top of the Bocksberg (2,380feet/726m) to the village at about 1,900 ft (580m). We were there for twelve days with quiet, often nearly still weather, ideal for any kind of winter sports. It was for Moira and me one of the best snow holidays of our lives, the rather gentle slopes being well suited to our fairly elementary ski techniques. Our main interest in the sport was always in the aesthetics of the snow scenery combined with mild satisfaction at the exercise in the sharp air. The temperatures ranged between about $0°$ C and $-16°$ to $-17°$ C, but because of the absence of wind and the well heated houses it never really felt cold. There was some more of the same kind of fun later that winter after our return to England, when the family were all out again on skis on Holmbury Hill in mid February.

There was a sequel to that Christmas visit to Germany the following summer, when our Catherine was due to finish at Göttingen and return to Norwich. It seemed that there was no other way to get all her stuff home, including some china which she had bought at a local pottery in the Weser valley, but for us to collect it all with our car. So we took a rather large diversion, and bought a few items for ourselves at the same pottery, before driving all her things round with us on our summer holiday in Denmark, southern Sweden and Norway, and so back to Guildford.

The journey to Göttingen gave us an opportunity to visit the frontier between the Soviet occupation zone and the western zones of Germany, which passed just a mile outside the town. The sharply cut, parallel-sided gash through the woods and fields in the valley was in stark contrast to the gentle rolling countryside. There were tangles of barbed wire and trenches marching straight through the fields and across the lane, where a notice baldly stated *"Zonengrenze"* (zonal frontier).The continual movement of military personnel and vehicles made it more than clear that there was no way through that frontier. We were to encounter the Iron Curtain again three years later on a deserted moor in north Norway, where the iron was decidedly rusty and no trenches were visible. But the iron fence was three metres or more high and the road that led through it was becoming overgrown with weeds and scrub even in the Arctic climate. There was a big notice in Norwegian forbidding photography. The notion of the mighty power in Moscow feeling threatened by the consequences of any of their people straying over the line and being contaminated by meeting some wily Norwegians was more than ridiculous.

It was in autumn 1971 that I first visited Lapland, for an international conference on Arctic climates over the last 10,000 years at Finland's northern university in Oulu on the Baltic coast, reviewing the evidence of surface geology, moraines, glaciers and ocean-bed materials, detecting past changes of prevailing winds and ocean currents. We later transferred to two field studies centres out in the wilds, first crossing the country eastwards by coach to Oulanka, in the birch and pine forest near the Russian frontier, and then back to Rovaniemi, the capital of Finnish Lapland, before flying north on 8th October to Ivalo and then on by bus, passing one arm of the lovely lake Inari, and farther north over the highest ground in Lapland to Kevo, near Utsjoki and the Norwegian frontier. It was a striking tour through alternations of forest, moor and lake, with brilliant orange colours developing in the long lake-side grasses as well as frost-heave features on the moors. These moors, covered with stunted birch and willow, as well as patches of heather and berry plants, had a distinctly end-of-season look, with the summer colour gone, but patches where the leaves of the blueberry and other berry plants were beginning to shine out with the brilliant scarlet red of autumn. The last day of this journey was made under a very clear sky with the sun, that greeted our arrival at Kevo in 70° N, just about to set at 4p.m. To reach the field study centre, in an appropriately designed group of wooden buildings in the forest edge, beside a small lake, on the shore opposite the bus road, we had a short boat crossing. By the next day there were signs of ice forming, and it was obvious that nothing short of a change of the weather bringing a south wind could stop the freeze-up at that latitude in mid October. We wondered whether there could be an awkward situation already developing for our departure on the 10th with too much ice for the boat but not enough to walk over. But this was in the event safely accomplished, several members of the conference taking a sauna in the special hut beside the lake the night before leaving and duly plunging into the lake after the heating. But I was happy to follow the example of a Danish friend who declared that, at 50, he was too old for that. We flew back from Ivalo to Oulu and then bussed along the 6 km road over flat land that had been at the bottom of the sea at the beginning of the century. This area was under some of the thickest ice in the last ice age, about three thousand metres (10,000 feet) thick at its height, and the rocks of the Earth's crust that bore that load are still rebounding, rising there at the rate of about one metre per century.

A month later, in mid November 1971, I was to have another cold experience, at a NATO meeting on all branches of North Sea science, held in the hotel in the sports centre Aviemore in the Scottish Highlands. I arrived on a Saturday by the night train and was met at the station by an old family friend. After visiting other friends and old haunts in the neighbourhood, on the Sunday afternoon we attended the little Church of Scotland kirk at Kincardine on the moor, and then I moved into the hotel. My bedroom in the modern concrete tower building, that breaches the forest and hill landscape, was not warm. The wind had begun to blow from the north, and the stainless steel frames of the sash windows, Victorian style, were evidently not really air-tight. There was nothing wrong with the hot water system, however, so I had a second hot bath, to warm up, each night about 3 a.m., and in that way secured a good night's sleep.

No doubt, the theory of holding a conference at the Aviemore centre in mid November was that the participants were effectively captives who could not escape the punishing work schedule, as there was nothing else in the area to do at that date. Lectures began at 9.30 a.m., after a good breakfast and, with a bit over an hour's break for lunch, a quick cup of tea in the afternoon, and a good dinner in the early evening, the sessions resumed from 9 till between 10.30 and 11 p.m. The supposition that there was nothing else to do in the area did not, however, apply to me. I escaped one evening to share another meal with our old family friend elsewhere in the centre. And on the Thursday afternoon, 18th November, I played truant and took my camera to photograph the, by then, snow-covered countryside, planning to take advantage of the snow to photograph one of the few points in Britain where the natural upper tree line is clearly shown by the trees higher up on the slope becoming gradually limited in height to stunted species less than one metre high. The uppermost limit of the trees hereabouts nowadays on a crag called Mhigeachaidh and continuing along the range nearer to Loch an Eilein is somewhat below 600 metres (roughly 2,000 feet), and at most points below 500 metres (1,640 feet), though the stumps in the peat, radiocarbon dated to 4,000 or rather more years ago, of many hardened, old, well-grown pines indicate that fine, big trees once grew higher up.

The snow had come on the 16th, and with a bitter northerly to north-easterly gale and showers and longer periods with snowfall, had continued to accumulate. On the steep hillside, consisting largely of snow-covered scree, with some larger stones and tree-roots, I did not succeed in getting to within fifty metres or so of the uppermost trees, though I took some satisfactory photographs and saw some deer. The light was failing about 3.30 to 4 p.m., when I got down to a more open tract, with gentler slopes, along the foot of the hill above Loch an Eilein, where a little loch (Gamha) lies. Some minutes later, when I had reached the main track around the lochs and was re-entering the forest, to my surprise I met four teen-age girls, probably only 14-year-olds, in duffle coats, going slowly up the track in the opposite direction. One of them approached me and asked could I tell them "the way to the bothy". This shelter that they were asking about was a good 2 kilometres (over a mile) ahead along tracks that are not always well marked, through the open ground not far from the foot of the steeper hill. The wind was strong in the open, about a force 7 gale, the temperature probably about -4° C. My response to the girl's question was "Should you be going out there, to the bothy, in this weather?" The reply was "Oh yes. We must. Our party is meeting there at 5 p.m." I was horrified and suggested it was dangerous. But they felt they had to go. I told them the way, in outline only, as it was impossible to describe the detail of the paths, and there was no sign of them having a map. As I made my way back to the conference hotel in Aviemore, I debated with myself whether I should inform the police. Ever since, I have blamed myself for not doing so.

First Tasks for Climatic Research

The temperature was -1° to -2° C all day in Aviemore, and conditions must have been terrible on the Cairngorm tops, probably around -12° to -14° and with a strong gale making it feel colder than that. And it turned out that that was the setting for one of the historic tragedies of the Cairngorms. The leader of the party that the girls I met were joining, herself a girl of 21 to 25 years, took them up on to the tops at the weekend, despite the weather. They got into difficulties and tried to build a snow shelter, but by the Sunday night nine of them were dead.

The winter, twenty-two years later, in which these lines have been written, has had some similar weather features, but has been more changeable in many countries and has seen an even higher death toll in accidents to people in the mountains. Among the suggestions as to what could be done about it, has been that everyone involved should make themselves thoroughly aware of the weather forecasts and that they should insure themselves so as to be able to pay the costs of rescue services. Proper awareness of the weather forecasts is an elementary necessity. But not everyone is equipped to understand fully the development of the weather situation or to recognise the significance of new developments and trends in their early stages. There is an obvious need for thorough training of people venturing on the mountains to cover these points. But there is no means of making insurance, or other recommended preparations, compulsory. One has to recognise that in the end there is no possible protection against folly - particularly in the leadership.

My wife and I were back in the Cairngorms in September 1976, at the end of an unusually hot summer. The heat wave continued there as elsewhere until 7th September, and a party of Belgian holiday-makers used the chair-lift to go up over 4,000 feet (1,220 metres) to the top of Cairn Gorm in just cotton singlets and shorts. They descended the far side of the mountain to the lonely Loch Avon, shut into its bare glen, at the foot of precipitous cliffs, which winds away for miles on the north-east side of the hills before any road reaches it. There they slept underneath the famous Shelter Stone at 750 metres above sea level, a massive boulder that has lain there since ice-age times. But, when they woke up on the 8th, they were surrounded by snow, which covered all the mountains in sight and down the northern sides to about 500 metres. It took all day for them to be rescued by helicopter, and taken to hospital in Inverness. This could, no doubt, properly, be regarded as an accident, but it was due to a cold front bringing Arctic air from the far north which must have been tracked and foreseen by the weather forecasters on duty.

For three days in October 1979 I was at another North Sea science conference, this time at a gracious Danish seventeenth century manor house, known as Sandbjerg Slot (castle), in the southern part of Jutland, overlooking the water of the Little Belt, which divides the island of Fyn from the mainland. The conference was concerned with how the climatic changes in the last thousand years have affected the cultural history of the community of the surrounding lands. This meeting had one very memorable aspect, due to the beauty of the place. The first day was fine and warm, despite the late date, with the sun shining brightly in a very clear sky, one of those calm, cloudless days that are commoner in spring and autumn than in summer. The meetings were held in an upstairs lounge with big, sash windows, which were open, looking out on a well kept lawn sloping down to the water. And the meeting had barely begun before a long-drawn-out succession of the tall, mostly three-masted sailing ships of yore began sailing majestically out through the sound to join the international parade on wider waters. It was nigh impossible for our audience to concentrate on the lectures, and even the speakers were seen to pause and turn round to admire the scene outside the open windows behind them. Nevertheless, the conference was an interesting one, tracing the development of the North Sea itself and its sediments and coasts, with particular attention to the last ice age and since, with the history of storms, the human history of the area, trade, fishing and cultural exchanges, shipping and ship-building. A rather similar conference, concentrating more on the low countries and coasts of the southern North Sea was held, mostly in the Flemish language, a year later under the leadership of Professor A. Verhulst in the University of Ghent, in Belgium. The book of that conference, published in 1980 by the local history department of the university, contains expert coverage of many sea-flood and sand-drift problems and their effects on the coast, notably by the Dutch scientist M.K.E. Gottschalk and others, supplementing the research publications of the Dutch Geological Service and in Geologie en Mijnbouw. The history of incidents of blowing sand is paralleled by experiences all along the low-lying coasts of western and northern Europe from Biscay to the Arctic. The examples include encroachments of the sea by flooding and crumbling cliffs and others associated with recession of the sea due to silting, and blown sand. Others again are due to land-rise along northern coasts with continuing rebound of the Earth's crust where it had been depressed by the great load of ice.

It is unfortunate that studies produced nowadays treat these and other matters related to changes of climate as if they are always, and only, attributable to the activities of Man and side-effects on the climate.

CHAPTER 20. HOLIDAYS

Since my return from the whaling voyage on a Norwegian-manned ship to the Antarctic, with Norwegian food and life and songs of the sea, I have had a hankering to return again and again to Norway, especially to the most unspoilt parts of the interior of the country where mankind and its pretensions are dwarfed by the scale of the landscape and its forces. But we have been equally held by the charms of the south and south-east coast, Sorlandet, with its low, rugged pine-clad hills, its sheltered bays and islands with the many little ports and harbours, where the old sea-dogs of three centuries have made their homes in the trim wooden houses whose white paint shines in the sun. It was there, on one of the offshore islands, where having tied up our boat, one sunny afternoon, while ambling casually round the island exploring, we were invited to join a party of the residents, sea-going folk, sitting in the open air outside their 200-year old white wooden house with its traditional red barn. We were hospitably introduced to Drambuie, the flavour of which gained some romance from the circumstances. The conversation was about seafaring exploits and the history of life there. There are some leafy nooks, with birches, maples and rowans, and little valleys too, on the islands and the coast and various types of old buildings, all wooden and well kept. We made our summer holidays thereabouts three or four times in each decade from the `fifties to the `seventies. This was by no means costly in those days and brought other advantages besides. Each time we took a different route for our journey there, choosing different ways over the mountains from the great west coast fjords, or coming from the east and north over Denmark and through Sweden and eastern Norway. This gave the young folk glimpses of other landscapes than the cosy south and south-east coast, viewing the wild mountain roads of inland Norway, with their glaciers and great waterfalls, as well as the historic wooden farms and villages of the interior, and meetings with occasional herds of goats, some sheep, and, more rarely, in the higher tracts some reindeer.

One virtue of the south-east coast of Norway, of Sorlandet, that remains almost unknown to British holiday-makers, is the prevailing warmth of the sea water in the summers, particularly in July. This is due to the light winds and sunshine that habitually play upon the shallow water in the bays and on the sheltered beaches, and on the rocks that have been characteristically smoothed to "whale-back" forms by the action of the glaciers of the ice ages. These features are even more marked in Oslofjord, where the average temperatures of air and water at the height of the summer - again especially in July - are a degree or so higher than in London or on

Holidays

European coasts north of the Bay of Biscay, though often nearly matched on parts of the Dutch, Danish and Baltic coasts.

The scenery of the south coast is enlivened by the warm shades of the pink and pale-grey granite of the often glacier-smoothed rocks and the mostly white painted wooden houses with their red crinkly-tiled roofs. Far in the interior some of the older upland farms and villages are largely unpainted, but everywhere there are also other buildings painted in the equally traditional style of nearly rust red wooden walls with white eaves. Smaller numbers favour other bright colours, such as a strong chrome yellow also with white eaves.

One fondly remembered, though rainy, holiday in 1951 on this coast, at Fie in Sandnesfjord, before our transfer to Malta, was really too far east to connect with the main ways over the mountains, but Norwegian friends of mine before the war had spoken glowingly of Fevik, with its bays and sandy coves, all with trees or woodland at their head, between Kristiansand near the south tip of Norway and Arendal. One of them, a shipping magnate's daughter, had been brought up in a great house in Fevik, which had since become a rather grand *pensjonat* (pension). So I wrote to the appropriate tourist office for brochures listing accommodation and followed up with letters to two or three of what seemed the likeliest houses in Fevik. We selected the one which sent the nicest letter in reply, interested and kind without being effusive. It was a good guide to character and introduced us to our friends of forty years, the Ramlets, who have been extraordinarily generous to us and with whom we have had many happy meetings in both countries. Helga Ramlet, the mother of the family, was a professional knitter of the now widely famous traditional, Norwegian (very hard-wearing) woolly garments, jerseys and cardigans. She supplied the shop (*Husfliden*) that is the official outlet for this clothing and other cultural wares over many years. We had many valued examples from her, sometimes as a gift, and some of them are still being worn by younger generations of our family. A fortnight before our eldest, Catherine, was to be married in 1975, their son, Ole (pronounced roughly Ōlie) Ramlet, then in his early twenties, arrived at our gate, completely without warning, on a rough farm lane in rural Norfolk which he had never seen before, bringing a wedding present. He was driving a hired camper-type van and brought his fiancee and her younger, teen-age, sister, Solveig, with him. (Incidentally, all the prospective marriages involved have since taken place and have turned out well.) Ole's father, Karl Ramlet, had emigrated to the United States as a young man and worked for some years not very far from New York, in a woodwork business, but then came back to the same area of south Norway that he had started out from, having acquired a knowledge of English and a valuable range of practical skills. Their son Ole is now a surveyor in charge of the county authority's office, covering a very wide area.

We approached the south coast with its idyllic little bays and islands by different routes each year that we went there: in 1956 by ferry from Denmark when, sad to say, the Skagerrak crossing made the children sea-sick; over land from Sweden in 1959; over the mountains from Bergen and the beautiful Hardanger fjord to Haukelisaeter and then down the long Setesdal valley with its ancient wooden farm buildings and great rivers in 1962; and by a longer way through the fjord country in 1964, from Bergen to Voss and Gudvangen on Naerofjord, a branch of the biggest fjord (Sognefjord), then over the heights of Sognefjell, more than 1,400 metres (4,600 feet) above the sea, to drive down the great central valley of Norway (Gudbrandsdal). This stage of the journey was done by a succession of camps with our car tent, after making a detour with the car up to a frozen lake nearly 6,000 feet (about 1,800 metres) up, which nestles just below the top of Galdhøppigen, Norway's highest mountain (2,469 metres - about 8,100 feet above sea level). In 1968 we were back again with our friends, the Ramlets, after camping in hot, sunny weather on Denmark's middle island, Fyn, then driving through Sweden and camping beside the great lake, Vättern, at Gränna, which will for ever be coupled in our memory with the "cream slices" - Napoleon kaker (Napoleon cakes), of flake pastry interleaved with vanilla-flavoured custard and cream alternately - already remembered from my first visit in 1935. The lake-side vegetation there had quite a tropical look about it in the warm, sunny weather.

It was lucky for us that these Scandinavian family holidays were taken at a time when the exchange rate was a good deal more favourable for travellers from Britain than it later became, and living there was relatively cheap, but we must also be considerably indebted to our good friends, the Ramlets, who seemed always happy to see us again. In return, we had Ole over to stay with us for some time in 1969, and there was a bigger family visit later on.

In August 1970 we were again briefly with the Ramlets in Fevik, in the course of our remarkable trek to Göttingen in Germany to visit our daughter Catherine and pick up her belongings at the end of her university course there. We went on to camp in the hills near Oslo, while I attended an Antarctic congress at the university and visited the Norwegian Polar Institute. We drove on from there into central Sweden to camp beside the lovely lake Siljan which lies, in a basin formed by a meteor's impact with the Earth some 360 million years ago. There were more visits in later years. And in 1980, the whole Ramlet family, Helga and Karl, Ole and Ole's wife Gunn-Elise, came over and stayed with us in Norfolk. Helga particularly fell in love with the gracious, contrived landscape around Ketteringham Hall, with its "fairy-tale wood", alongside where we were then living, and through which the lane wound its way on. Until late in 1974 we had had red squirrels in that wood, but, as elsewhere in Norfolk, they had disappeared and been replaced within one week in November of that year by a family of grey squirrels living up the same tree. The greys were visibly more vigorous and less ready to hibernate than the last of the reds, though there were reported still to be red squirrels among the pines in Thetford forest many years later. Conservationists in much of England who treasure our native red squirrels are not being quite consistent when they join in the unbalanced campaign against pines. Like so many attitudes that run to unreasoned extremes, this is largely a matter of fashion, one that has spread like wild fire just over the last ten or twenty years. There are also hints that this fashion draws its strength from a sort of nationalistic prejudice against anything regarded as alien to the lush southern English countryside of Jane Austen's Hampshire scenery and to the grassy moorlands of the Pennines. But this ignores all those areas where pines are a well-loved part of the English scene, as in Ashdown Forest in Kent and in many parts of East Anglia as well as around Bournemouth and parts of Dorset and Herefordshire. I was shocked to hear of a ramble, organised by the National Trust, no less, in north Norfolk a few years ago, on which the walkers were told at the outset "And if you see any seedling pines on your way, just pull them up" - surely an unwarranted interference with nature in a part of England where pine pollen seems to have been an element, even if a small one, in the pollens deposited in all ages in the last ten thousand years! In this writer's experience, in many parts of England, Scotland, Scandinavia, and even in Ireland, there are few pleasanter ways of idling than to lie back on a bright, sunny day under a few pine trees or in an open, natural pinewood, with its carpet of heather and the small berry plants, and look up at the clear blue sky between the tree-tops.

We learnt a lot from our Scandinavian holidays about life in small communities, albeit communities geared to the use of hydro-electricity and the advantages of electric bakeries etc., etc., as well as about navigating on foot and by car in the wilds in terrain where one must never take unwise chances with nature - with the weather or the scale of the landscape - where one cannot assume that whatever mistakes one makes, or whatever difficulties one gets oneself into, nature is basically friendly. In the wild and lonely places it is more obvious that human beings must help one another. And one sees international affairs from a somewhat different, but very humane, perspective. There is a discipline experienced by those who live in such places, which tends to have a benign influence on the side of reasonableness, though the case is different for those who live in the poorest and really isolated districts of the far North, where there are fewer natural resources.

It was thanks to Ole's boat, and the skill he developed with it, that we were able to explore the offshore islands and skerries, and later on he and his young family also had a friendly horse and an old carriage that took us through the leafy lanes of the bigger islands that are nowadays connected to the mainland by bridges.

There is a picturesque little old steepled church, set at a visually strategic point on the roadside between Fevik and the little coastal town of Grimstad, overlooking an inlet where ship repairs and some ship-building are carried on. Inside this church at Fjaere there hangs a splendid model of an old three-masted ship, suspended just before the chancel, as is traditional at many places in Norway. In Grimstad, at the same apothecary's shop in the middle of the main street, where Henrik Ibsen worked as a young man, we used to go for remedies for our family ailments such as hay fever and cold sores. Grimstad was a convenient little place with its entire shopping centre in an area scarcely three hundred metres across and good go-ahead shops that could provide excellent anoraks, mackintoshes, and gum-boots, as well as rucksacks, swimming things, photographic films and medicines, compasses and thermometers, and most kinds of instrument relating to weather, climate, and outdoor problems and pursuits. It is also a picturesque and zestful little town, with its smartly painted, mostly white, wooden buildings and red-tiled roofs, sparkling in the sun, beside the water and the boat harbour that penetrates to within a few yards of the centre. The town is dominated by its slender church spire, which towers above the place and its harbour from the slope of a hill on the edge. In Arendal, the bigger neighbouring town to Grimstad, the other side of Fevik, there is a first class book shop (Danielsen's), also good for maps, as well as other shops well stocked with goods ranging from practical walking, sailing and camping wares to pocket magnifying glasses, traditional Norwegian designs of china and wooden plates, and of course woollen knitwear. It is to be hoped that Grimstad will keep its charming, and practical, character in the face of the more extreme unwelcome pressures of modernisation better than the Danish port of Esbjerg, where the introduction of American-style hypermarkets around the outskirts has caused the disappearance from the town centre of many useful small businesses. One example we noticed there was the removal from the inner town of the versatile shop of the Cykkelkompaniet (the so-called cycle company), which sold not only bicycles but a wide range of elegant Danish stainless steel wares, from cutlery and egg-cups to dishes of many sizes and descriptions, including sauce boats and much else.

Another character, whom we fondly remember from those holidays at Fevik Pensjonat is Asbjørn Bakke, a schoolmaster from Oslo, who was a regular summer visitor there and already a firm friend of the Ramlets. He was always full of fun both in the house and on the many sunlit bathing beaches among the pines near Fevik. He always greeted us with torrents of high-spirited fast talk in both English and Norwegian. He was a devotee of football in both countries. A short, fair-haired, freckle-faced man, he was nearly always his cheery, smiling self, till sadly around 1980 he failed rather quickly and soon died. The water on those beaches and in the sandy coves at the foot of the rugged little birch and pine-clad hills was usually in summer surprisingly warm for bathing, often up to $20°$ or $22°$. But if the wind became strong from the north or north-west, the warm water could be blown out to sea and replaced by cold water fetched up from the bottom. At such times the water temperature on the sands could be as low as $14°$, even in July and August.

But although our memories of that coast are nearly all of warm sunshine and good friends, it can be seriously stormy at times in the winters. I took the opportunity to visit the local archives in Kristiansand and Stavanger, where I found details of some severe storms and the ships' weather reports, which I later used to map the storms, concerned in a study of historic storms that was published by Cambridge University Press. Another personal link with that coast was forged later, by a strange chance, when we were at home living in a tiny Norfolk village. There was a Norwegian meteorologist, Harald Christensen, who, acting in 1942 on a tip from a senior colleague in the Norwegian Meteorological Institute in Oslo under the German occupation during the war, had escaped on foot through Sweden, and thence by air, to Britain to work in the Met. Office until 1945. He had married a British wife from Luton in those years, and she went back to Norway with him, had learnt the language, and lived in Norway ever since. They had bought a cottage in England, in which they stayed from time to time, at Hethersett, near Norwich, only two English miles from our home in Ketteringham. I had heard tell many years before of the wartime romance of "this fantastically good-looking young couple", as the Met. Office assistant girls who had known them at the chief weather forecasting office in Dunstable described them to me, just after they had left for Norway in 1945. But we ourselves had never met them until our wives met accidentally at a Women's Institute evening meeting in Norfolk in 1973.

One of Harald's best tales was of how the British managed to break the Nazi secret code for weather messages during the war. This was in fact made easy because of a daily upper air balloon ascent, which registered temperatures, pressures, winds etc. at a great many heights somewhere in

General Franco's Spain. The results were transmitted as long messages by radio in clear (plain language) by a local station for Spanish use and in the day's code for the Nazis' use. All that was needed to lay bare the code was to compare the two transmissions!

We soon became firm friends, and Moira and I enjoyed many an evening with Harald and Ruby at their cottage in Hethersett, where Harald used to make special delicacies for us, such as *Gravad Laks* (a form of smoked salmon), before it became readily available in England, and Wienerbrød (the authentic Danish pastry, - quite unlike the conventional, heavy, jam-filled English copies), and excellent home-made white rolls. He and Ruby used to come to us too, and we stayed with them several times in Norway, either in their hut in the mountains in Valdres, in central Norway, or in their retirement home on the south coast near the historic port of Tonsberg. By a happy coincidence, their Tonsberg cottage was right beside the sheltered water where the whaling factory ship *Balaena*, with me on board, had lain for three days in September 1946 when we were picking up the whaling crew. Our ship was not 200 metres from where we now came to stay with Harald and Ruby Christensen.

Our family holidays in other years were quite varied but with an emphasis on long journeys to this or that region of interest, and getting to know what the life of the people there was like, by returning for a few more visits. It has always seemed to me preferable to get to know a few well-liked regions well rather than become an endless, and maybe ultimately aimless, globe-trotter. However, we did go seventeen or eighteen times, generally twice each year, to the same ancient farm, on a hill-top in well wooded country in Herefordshire. We went for short breaks at Easter and the October half-term holiday during our youngsters' childhood. The house was a fine old half-timbered building, reputedly dating from about 1450, with grand old dark oak stairs set around a square stair-well. The connexion began through revisiting areas associated with my uncle, Reginald Brierley, my mother's brother who was county architect in Hereford for many years from the end of the first world war until his sudden death of a heart attack in 1932. He was unquestionably my favourite uncle, who really understood my difficulties with my father, and whose visits to our home always had a soothing effect. He and my aunt, Louey, had driven me, mostly in the dickey seat of their little Rover 8-horse power car, all over Herefordshire and on into central Wales, on his rounds inspecting schools and other buildings in the nineteen-twenties. That had given me my introduction to many interesting places, particularly memorable being the remains of Llanthony Abbey, which still visibly dominate that deep valley among the Black Mountains on the Welsh border and clearly had provided a protective haven of civilization and caring in that remote place before the Reformation. We were very well looked after by Grace Warren and her husband, Sam, at Joanshill Farm, Checkley on those holidays, including one time when one of our girls had German measles while there. We enjoyed many a climb and ramble in the wooded hills near by. Our children had great play on an old, abandoned horse-cart, which became a "pirate ship" in the unlikely surroundings of the orchard. The old farmhouse had an overhang of its upper storey, beyond the walls of the ground floor, and on more than one occasion the snow used to come upwards, blown into our bedroom, between the floorboards. That was "mildly" unfortunate, because snow fell at one time or another during as many as nine of our spring visits. But it was never more than a joking matter, as no great quantity of snow got in and there was no real discomfort caused.

Later on, in 1965, we bought ourselves a cottage beside a hill track on the edge of a hamlet called Westhope, ten miles out of Hereford to the north-west on the way to Wales, and had it for eighteen years. We made many good friends among the country dwellers in the area during that time, most particularly the builder, Geoff Norman, who put the cottage in order for us and his wife, Ruby, and their family. Geoff also looked after the property in such matters as turning the water off and on again at the beginning and end of each winter. One autumn he was called out to clear away a sizeable landslip, when the bank at the edge of our field fell away and blocked the steep public lane leading to most of the cottages on the hill. We also had stalwart support from Mr. and Mrs. Reynolds, Will and Rhoda, who lived more or less next door, just down the hill from our cottage. We allowed Will to graze his sheep in our field. There was a round of seasonal activities for us in and around Westhope such as gathering damsons off the trees in our field and brambling in the autumn. The four old apple trees were on their last legs and did not produce much. But there were also many lovely places for picnics and for hill walking, mostly in central and south Wales. On one great day out on the Brecon Beacons, heavy rain set in on our way down from the highest top, so heavy that the children's clothing was completely soaked long before we reached the car. We had to drive them back to Westhope naked under the rug in the car with the heater on full. However, a good hot meal in a hotel in Leominster after our return put them to rights. The greatest expedition we managed from the cottage was a climb to the top of Cader Idris, one of the higher mountains in North Wales, on an excellent day in bright, clear weather.

Having a cottage, as a second home, was an experience that had drawbacks as well as its good points. There was a lot of effort involved, particularly for my wife, in getting everything ready for the journey and

Holidays

During and since my time as head of the Climatic Research Unit at the university in Norwich there have been opportunities of combining pleasure with business, for instance in organising the recreational excursions at times when we were running scientific conferences. My wife played a valued part in these activities. There were also moments of amusement, as when one of the Unit's scientific staff, who was about thirty years old at the time, was on duty one Sunday afternoon welcoming our foreign guests arriving for a conference and showing them to their rooms on the university campus. When he offered to help a Greek professor, a well-built man, and his wife, by carrying one of their bigger cases, the professor waved him away, saying "Oh No, no, no. My wife is strong"!

As the years went by, and we got older, also we have often extended our Norwegian holidays by adding some time spent in Sweden as well. With our increasing age, and some physical handicaps developing, the gentler landscapes of Sweden have proved increasingly appealing. And the usually sunny Swedish summer climate adds to the charm. Mostly we have explored the central districts of Sweden from Värmland and Dalarna in the west all the way to the Baltic coast, where there are old fishing settlements of great beauty. The average sunshine hours across this belt are generally greater than anywhere else in Europe north of the Alps. It happened that when I was working in the Harrow Met. Office and we were living in Radlett in the late nineteen-fifties, we found ourselves living quite near an old Quaker friend of mine from pre-war days, Martin Lidbetter, and the Swedish wife he had met and married as a result of his work with the Friends' Ambulance Unit during the Finland winter war of 1940. It was a happy renewal of contact, and his wife Eva advised us in making our holiday plans many years later about the countryside and coast near her old home in eastern Sweden. When she and Martin were living in Stanmore, Middlesex, we had one unforgettable experience on an evening when we had gone to have an evening meal with them. They had been back in Sweden the previous summer, and Eva had developed quite a longing to taste again a traditional northern Swedish delicacy, called *surströmming*. It is a sort of herring that penetrates right into the northern rivers during the summer time. When caught, the fish are allowed to rot slightly, then buried and afterwards pickled. Our friends had driven to one village store after another rather late in the summer, till finally they found just one tin of this delicacy that was somewhat past its "use by" date. And we were to be honoured to share this tin with them as a starter to the meal. However, as the tin was slightly domed, Martin and I went out into the garden to open it. And when we did puncture the tin, a jet of bubbling liquid shot up vertically into the air. The stench was not good, so we moved the offensive object from time to time from near one neighbour's hedge to another, while we discussed what to do with the contents of the tin. But Eva was still so keen to have the dish that the contents were transferred to our plates and served to each of us. Eva had the biggest portion and straight away stuck into eating it with enthusiasm. The other three of us just sat and looked apprehensively at the plates before us. Martin and I each ate two or three small fork-fulls, but Moira did not really venture any. Finally, we all gave up, but Eva finished her helping. There were no alarming effects that evening, but I telephoned Eva anxiously next day to make sure that she was still none the worse. I have to confess that we were each so preoccupied with our own fears of the possible consequences that I have retained no impression whatever of the taste.

  We had one more Swedish holiday that has etched itself deeply on our memories, partly because it was one of our very best winter holidays, with the beauty of the scenery and the weather, and partly because of the risks taken by some of the younger members of our party. On 19th January 1979 Moira and I set out on the D.F.D.S. ferry from Felixstowe to Gothenburg with three small cars and six girls, including our daughter, Kirsten, making up the party. All but one of the girls were medical students. The weather was cold, and we drove to Felixstowe through falling snow. There was pancake ice on the sea surface in and near the Kattegat as we approached Gothenburg on the 20th, but in Sweden the weather was fine and the roads clear of snow and dry. We were driving under a cloudless sky, with a late moon. Before the ship reached port at 1 p.m., I had carefully briefed the other drivers in the party about the road journey, stressing the desirability of the three cars keeping within sight of each other. However, one of them was very much in love at the time and had lately got engaged to be married, and I could see that she was not really giving her mind to the briefing. By the time we stopped for petrol at the farther edge of Gothenburg, we had already lost sight of her, and we never saw her and her one passenger again on that journey. Our anxiety grew, as other members of the party expressed doubt whether Janet really knew the name of the village, Ekshärad, that we were making for, 215 English miles inland, in a northerly direction, from Gothenburg. As it was soon dark, there seemed to be no better plan for the rest of us than to go on with our two cars to our clearly stated objective and rely on our wanderer getting in touch sooner or later. Our instructions were to go to the garage in the middle of Ekshärad to collect the key of the hut we were hiring and further instructions. When we had put about three quarters of the journey behind us, we stopped the two cars on a straight stretch of the road, under the pines and at a place where the ground sloped quite steeply down to the river. There, under the bright moon and in the frost, we ate our ready-made picnic, which we had from the ship. The other girls told me then that Janet had not had her car, which was an old Mini, serviced for the journey. Our worries about her and her passenger were not lessened by that. We reached Ekshärad a little before 8 p.m. and collected the key. Still no sign of the wanderers, who - we imagined at that stage - had gone on ahead somewhere or other on the wrong road. We decided to post one member of the party at the garage, taking turns, to wait for them. The night was fine, clear and still, but the temperature was -8°. At half past ten, our lost ones turned up. Janet and her friend told a terrible tale of trouble with the car, which was leaking petrol and having continually to take on more. It would only go at about 15 to 30 miles an hour, which with repeated stops for more petrol meant very slow progress. We never did discover just where they had been, but evidently not far from the direct route. We sustained a further shock when we discovered that the neatly folded rug, which was the only thing in their car boot, was saturated with petrol. It was lucky that neither Janet nor her passenger were smokers! Part of their week had to be given up to getting the petrol tank attended to - so far as the garage in a remote country village in Sweden could arrange that.

We had a week of mostly calm and beautiful weather with the frost slowly intensifying. The skiing was gentle but good, on wide open tracks among the rather small birches and pines. I remember particularly one or two lovely, still days, when the low sunlight caught the occasional tiny ice crystals that were floating about lazily in the cold air and made them sparkle intermittently as minute flashes of brightness here and there all about us. There was a steeper, bigger hill rising above the trees near the holiday village, and it provided some fast going on beaten down snow. There were also a few trails marked out by distinctively coloured ribbons every few yards along the way. Our own daughter provided another cause of alarm one afternoon, when she set out, leading several of the girls, intending to follow the shortest, 1.5 km. trail, but mistook the colour marking and in fact led them along the longest, 15 km trail starting late, almost as dusk was beginning to come on. Happily, they completed their round safely, without harm, though our anxiety had been considerable as the sky was very clear and the temperature falling lower than before.

By the end of our stay, on the 27th, the temperature fell to -30° in the valley, -23° on the higher ground in the holiday village. And the cars which had been standing in the open were very difficult to start, because the oil in the engines had become so very stiff. Luckily there was a 2 kilometres long slope downhill all the way from the holiday village to the valley. So, after our car had been pushed by many willing hands to a suitable starting point, we were then pushed on more than a kilometre in third gear down the hill until finally the engine fired, and we were on our way. We had just over seven hours to get ourselves to the port in Gothenburg. And this was safely accomplished. But there was one catch-point on the way. We came beside a heavily iced up bit of canal with a heavily iced small vessel on it. It looked so blocked that we joked that it would be a long time before that got moving. But a few minutes later we were driving up the long ramp to where the broad road went over a modern bridge, when just as we passed a red light began flashing, and Kirsten in the back seat of our car was the first to realise that was the warning that the bridge was about to open. She said "for heavens sake, drive: drive fast, the bridge is about to open". We got over, with our hearts in our mouths. And the bridge then opened, and as soon as we had gone down the other side we looked round and saw the roadway behind us tipping up into the vertical position. And there was the badly iced-up little ship making its way through. When we had recovered from that and reached Gothenburg, we started worrying about Janet again. But she arrived with just half an hour to spare; though, after arrival in England, at Harwich, her car stopped again just one mile along the road and never moved again!

We were home in the afternoon of the 28th. The air over the sea had felt very mild after Ekshärad, but the snow scene at Felixstowe was not much different from Sweden as we left it. And the ground around our home was not clear of snow until after a whole day and night of rain six days later.

CHAPTER 21. NEW DIRECTIONS

Since my retirement from the directorship of the Climatic Research Unit there have been changes there and in the direction of my own efforts. My immediate successor, Professor Tom Wigley, was chiefly interested in the prospects of world climates being changed as a result of human activities, primarily through the burning up of wood, coal, oil and gas reserves, and by changing the face of the Earth by the destruction of forests, draining of marshes, and increasing drying out of the soils elsewhere, but the creation also of great new water bodies (reservoirs and enlarged lakes), and the prospects and impacts of global warming. After only a few years almost all the work on historical reconstruction of past climate and weather situations, which had first made the Unit well known, was abandoned. There was an exception in the case of tree-ring studies. Keith Briffa and his wife, Sarah Raper, took this up. Tom Wigley and his wife, Astrid Ogilvie, have continued their tree-ring work on northern Scandinavia and Finland, and the latter has continued her historical researches in Iceland.

There is much interest these days in the possibility of future development of climatic change due to Man's activities and it is now widely thought that the undoubted warming of world climate during the twentieth century is attributable to the increased concentration in the atmosphere of so-called greenhouse gases, meaning gases (such as carbon dioxide and some oxides of nitrogen) whose effect on the balance of incoming and outgoing radiation is not neutral but traps some of the radiation received. This concept provides a good explanation of some periods - as between about 1890 and 1935 and around 1987-92 - of rising world temperatures but does not readily account for certain periods of over-all cooling, some of which are associated with largely unexplained great increases in the outflow of cold water from the Arctic seas in or near the Greenland-Iceland sector. A new international research effort funded by the European Commission is to be aimed at assembling the relevant facts on this, focusing particularly on the early instrumental observation period 1780 to 1860 and the Late Maunder Minimum period 1675 to 1715 in the Atlantic Ocean, reconstructing daily and monthly mean situations, using a general circulation model with various sea surface temperatures and exploring both heavy and light sea ice cover situations. But, because mathematical models can so easily be designed to fit any theoretical concept, it is difficult to obtain fully convincing results by these methods. The coefficients in the equations which represent any empirically found relationships between causes and effects can be changed at will to simulate results that may be convincing enough, and seem to express satisfactorily valid relationships, and yet after a few years turn out not to be a lasting expression of reality.

One dubious practice, which seems to have been gaining ground in recent years, concerns innovations. There is no doubt about the verb to innovate, which is characteristically spoken with most emphasis on the first and third syllables. But there seems to be a cult growing up among the people most included to innovate - and particularly among those who like to be known as innovative - of transferring the stress to the second syllable or diminishing the stress greatly and pronouncing this word quickly with little, or only quite colourless, distinction between the syllables. This may be intended to convey the impression of some superior community that is always innovating, but the most obvious result is to obscure the origin and hence the meaning of the word and make it quite hard even to hear clearly - an automatic, and fitting, reward for attempted pomposity!

I have therefore found it preferable to continue to follow my own interest in climate and social history and its implications for war and peace and human welfare. Along with that, I have felt called to lend a hand where I could with the social and political problems of our own day, that are particularly acute in Britain. The country's difficulties seem to be at least partly, and maybe largely, due to loss of the accustomed, and traditional, sources of profit and income in the old British empire and the supposed special relationship with the United States of America. The effects have been sharpened by the prevailing inability to take a realistic view of these things. The superiority which Britain formerly enjoyed in manufacturing, financial services, and trade, has by now largely passed into history. And the sense of being one community with shared hopes and responsibilities has withered and been almost killed by the scorn of the country's leading politicians in the nineteen-eighties and since. This was perhaps the logical end-product of the class-war and industrial strife and had been growing and festering for some decades, until the scorn has become mutual; an in-built feature of Britain's divided society, made worse by government policies that are widely seen to have been opening wider the gap between rich and poor, and the welfare state itself has been progressively distorted to give greater and greater advantages to those who can make independent financial provision for their health and old age and the education of their young.

There seems to be just one respect in which the counter-attack by the political right in Britain since the late nineteen-seventies can be seen as justified from nearly everybody's point of view, namely in its criticism of socialist uniformity as an ideal. Granted that uniformity makes for a tidy country, many people are sure to complain that the results are dull. But like most thinking people of my generation, and especially those with a Christian background and beliefs, I was naturally interested in socialism, particularly in the efforts to establish a genuine social democracy as in Scandinavia (and in Germany before the Nazis). Two of our best friends were German social democratic refugees, Berliners, who escaped to Britain in the last years before the Second World War. Willi Derkow had been a trade union leader in Berlin until 1935 and was to broadcast in the BBC's German service after coming over to England. His wife Anne-Marie Derkow was a school teacher, who taught our girls German in Godalming County Grammar School. They and old Pastor Mensching undoubtedly affected our thinking on social matters, but the strongest influence has probably been our admiration of the calmer state of political life in most of the Scandinavian countries than in ours. It is widely accepted that the political differences between the opposing parties there are less than in Britain. After forty years of Social Democratic governments in Sweden, the change-over to a non-Socialist government was achieved calmly, and when after a few more years the Social Democratic party again acquired a majority in the parliament, the change back again passed quietly and without turning everything upside down. And in Norway, also after many years of Social Democratic rule, when a Conservative government had its small majority reduced to zero, the prime minister himself suggested that the Social Democrats should take over the administration. To us, the most enviable lesson of these events is the acceptance that different parties can make compromises and can even work together for the common good.

My personal experiences in Britain of the difficulties, amounting to near impossibility, of securing the funding that the newly established Climatic Research Unit required to sustain its life and support the fundamental studies on its programme opened my eyes to the vulnerability of new ventures to any sort of dictatorship, whether from a completely government-funded socialist state or from over-mighty individuals occupying strategic positions of power in a small to medium-sized country. It was the freedom offered by well resourced independent bodies - ranging from wealthy individuals to research foundations - that made it possible for the Climatic Research Unit to survive.

I have therefore come to see most hope of a change for the better in Britain in the development of a more freely democratic, "middle of the road" party - something like the Liberal Democrats - and a reformed constitutional structure that should take some of the heat out of political exchanges. By good fortune, our whole family including, so far, all the spouses and children, are of the same mind on this. We particularly regret all the confrontations perpetually engendered by the "first-past-the-post system" of electing members of parliament and, particularly, the evident result of this system in attracting into British political life far too many who go in for it chiefly for love of the battle. This system has made it almost inconceivable to most people that people from different parties might agree policy on practical grounds, to enact this or that measure, rather than decide all issues by party dogma. This system unquestionably makes Britain's troubles sharper and the difficulties worse. It stands in the way of reasonable debate, aimed simply at finding the best solution to any problem and foreseeing snags. But, clearly, such rational problem-solving does not make a strong appeal to the more primitive instincts of seizing party advantage and scoring points off any opposition. These ways of doing things separate us more and more from our partners in Europe and encourage the continuance of old habits such as the reliance almost exclusively on America and the British Commonwealth countries for all ideas and notions of how to organise things. It must also encourage a comforting acceptance of all our (sometimes outworn) customs and designs. The predominance of the English language certainly plays a part in this too.

Surely, the most sensible course in today's world is to develop a fairer and more open democracy, to glean good ideas from wherever they may come, and to organise the taking of law-making decisions as appropriately as possible: items of only local concern to be decided in local or regional councils or parliaments, items that apply to a whole nation to be decided in the national parliament, and things of still wider relevance to be enacted by the European parliament or an even wider international body. Something like this is surely required if the laws and regulations enacted are to be both practical and fair.

EPILOGUE - Language never stands still

It is appropriate in a work such as this to consider with some care the language that we use to communicate the lessons of experience and the outcome of our scientific studies.

Over many years, weather forecasters and, in fact, everybody trained in the Meteorological Office in Britain for work in any branch were strictly instructed to avoid using words that express personal or emotional reactions to the weather foreseen or to other aspects of the physical environment. But this, like most rules in the teaching of English, seems nowadays to have been abandoned. This change may have been introduced as early as the nineteen-sixties. Almost daily we are told instead what the "best temperatures" are expected to be, regardless of the fact that what is best for one person or activity, for one purpose, one animal, one crop, or another, is not necessarily the best for all others and may even be unlikely to be so. What our forecasters usually seem to mean by this slipshod usage is just the warmest or highest temperature to be expected. This is in line with the cult that has been growing up in Britain of regarding warmth as more important than anything else. Hence, any reference to snow in forecasts is habitually made in heavy, doom-laden tones, clearly implying that such an occurrence is regarded almost as an outrage, at least in comfortable southern England.

In reality, much more frequent intrusions of severe weather, into our modern sheltered lives in most parts of this country come from gales and violent storms of one kind or another and, in some districts, from thunder and lightning, or from floods and droughts. Formerly, even from Roman times, this country was famous for its mists and fogs, and from the first industrial revolution onwards some of these fogs became very dense, and poisonously polluted, "pea-soupers", as they were called (because of the colour of dried peas) in which at their worst one might not be able to see more than a yard or two. But these conditions of yesteryear have become far less common than they used to be, thanks in great measure to the Clean Air Act of 1956 and too the drive for more efficient fuel burning in industry and in domestic hearths. Nevertheless, very short visibilities are still commonly encountered on cold nights, in damp places, and on high ground, and are sometimes entered abruptly when travelling. They often cause deadly pile-ups of smashed vehicles on motorways. Almost equally dangerous are the often very local patches of ice - sometimes clear, almost invisible, ice - which can form very quickly under clear skies (and clearing skies) at night, and even after the sun comes up in the winter time, or whenever the sun is hidden by cloud or is very low in the sky.

Epilogue - Language never stands still

These hazards are often well covered nowadays in radio and television weather forecasting. But what is one to make of forecasts of "reasonable temperatures"? This is a prediction that some British forecasters frequently make. (Maybe the word should be "moderate", "middling", or just "average" temperatures.) Great care is needed in wording forecasts and warnings to be realistic and unambiguous.

Nowhere has the development of language been better explained than in Professor Otto Jespersen's books, notably in his "*Language, its nature development and origin*", published in London in 1922 by George Allen and Unwin. It had reached its tenth printing by the 1950s. An important feature of the book is the author's special attention to the times and circumstances in which rapid changes in languages occur. Jespersen's unusual breadth of knowledge enabled him to illustrate his thinking on this by examples from languages and dialects in many countries all over the world and instances from the time of breakdown of the Roman Empire to the appearance of those modern languages which are derived from Latin to the language of children in other places in our own times. The common characteristics of periods of rapid change and new development seem to be circumstances of breakdown and chaos, due to wars, revolutions, and pestilences, leading to the disruption of teaching and of the authority of both teachers and parents. We have seen these conditions - or, at least, some of their features - in Britain and other countries in the latter half of the twentieth century and their consequences in rapidly growing ignorance of the language and its usage. But, in spite of all this, some things go on as before. Most notable is the continuing appeal of simplicity, clarity and directness. So it is a common experience - as Churchill knew and emphasised - that those lecturers and writers in England today are most widely and readily understood who use the simplest, most direct English, preferring the shorter words of Anglo-Saxon and Scandinavian origin wherever available. That has been my own experience all through my career as it was my grandfather's experience a century ago in all his teaching and writing. And it often brings openly expressed thanks and appreciation of the clarity gained. Too many people love to use long words, which they think sound, and look, learned, but only succeed in making themselves harder to understand.

So, what has been happening to the English language during these last eighty years or so?

Looking back to the school classes of my childhood in 1919 and thereabouts, I am continually amazed by how much the language has changed during my lifetime, particularly by how many rules, and how many interesting quirks and by-ways of the language have been abandoned and soon forgotten. In the quicker pace of life towards the new millennium, it seems there is no time for quaint and nostalgic forms of speech or writing, whatever resonances it has and even when, as in some cases, the older forms were shorter. My mother often used to say "I never dreamt of that", whereas nowadays the past tense of dream is mostly spelt the American way as "dreamed". The pronunciation then becomes changed accordingly. But it may be that the changes in these last seventy to eighty years are hardly more than a tenth of the changes in the prevailing speech in this part of the world over the last seven to eight hundred years.

There are some other changes, besides the examples just quoted, which have a more insidious effect, multiplying the changes that follow. The noticeable preference in late twentieth century England for spelling words of Greek origin with a C rather than a K, as Greek does it and as was the custom here in the not so distant past, has led quite quickly to a change of pronunciation and then to misunderstandings. For this reason, the Americans seem to have been wise to retain the K spelling of the words "skeptic" and "skeptical". In those cases, the English spelling with a C is beginning in the nineteen-nineties to have some strange consequences, leading to the words being pronounced - even by some members of parliament and government ministers who like to sound very modern and go-ahead - as "septic" and so on, thus losing the distinction in the spoken language from talk of septic matter and infections and the problems connected therewith!

Of course, languages cannot be expected to stand still and completely unchanging as time wears on. The urge to achieve freshness by saying things in a new way is often a welcome and useful one. But there are also less attractive motives that keep the impetus for change going, such as pomposity and the wish to sound erudite even more than to be understood. And there will always be "copy-cats" among those who lack originality, those who are quick to identify "a trend" and copy it, in order to be "with it" and "in the swim" and who particularly ape whoever their popular heroes of the moment may be. The latest fashions that are perpetually taken up by the trendier public speakers reveal what surprisingly small vocabularies many have at their command and so have to fall back on their readiness to copy.

A notable example of these sudden fashions for a new turn of speech started in 1992 - I think that was the year - when our British politicians, no doubt under the dreaded foreign influences spreading from the European Community, or "the Common Market" as so many of our Euro-haters prefer to call it, suddenly noticed the French usage of the word to "address". (This was after the Thatcher years, in which no patriotic Tory politician would admit that Britain could possibly learn anything from another country.) Soon every last one of them was "addressing" all the awkward problems that came up instead of getting on with tackling them. After all, it is much easier, and far more immediately beneficial to one's image, to talk in a grand sounding way about a problem than to do anything about it! The years seem to bring an endless succession of such fads and foibles among our public speakers.

In the nineteen-seventies, they were all busy making opportunities to say "There's no way I'm going to do that". Endless roundabout opportunities were found of introducing the words "*No way*", usually with heavy emphasis. More recently, "*currently*" has become an "*in*" word. Hardly any speaker likes to say "*now*" any more, and alternatives such as "*at present*" and "*presently*" seem to have been quite forgotten. Similarly, the words "*altogether*", "*completely*" and "*utterly*", evidently never come to mind when one can say "*totally*". And the useful little verb "*to damp*" or "*damp down*" any excessive outburst of anything seems to be completely forgotten and is almost always confused with "*dampen*". This has reached the point where one of the Meteorological Office weather forecasters on the BBC has once or twice been heard to say to the listeners that "the shower activity will be dampened" later in the day, meaning that it will dry up!

It used to be said in England jokingly that Americans will never use a one-syllable word if there is another word with three or four syllables that will do. Hence the popularity of words like "*utilize*", "*volition*" and "*executive residences*" and so on. But these examples show us that the same disease is rampant in England now. Instead of using these circumlocutions, we should have the courage to try our "*own free will*" once again and be proud to live in our own houses and homes, even if they are mortgaged or rented! So too it has lately become fashionable to talk of house prices "*reducing*" (reducing what?, one wonders), when they are simply falling.

Apart from these examples, some present-day fashions in speaking involve no more and no less than using the wrong word. Again and again nowadays official speakers concern themselves with whether something is "*accurate*" when accuracy is not in question, but "*reliability*" is what it is all about. There was yet another discussion on the radio half an hour after this was written as to whether some risk assessment for insurance purposes had been accurate, because the figure was being disputed by an aggrieved party. But no risk assessment can be accurate - nor can it be inaccurate - before the event. The best that an insured person can ask is that the assessment was responsibly made by trustworthy people, using some open procedure and that it is therefore as *reliable* as they can make it. No doubt, speakers who use the word accurate in this connexion do so because they think it sounds scientific.

Epilogue - Language never stands still

In British schools there has by now been such a long-standing preference among the pupils and would-be university students for arts subjects, despite an uneasy suspicion that science studies might be more useful, that too little attention has been given to the strict meanings of words, particularly any that convey a scientific concept. There are obvious signs, too, that words and expressions are picked up from radio and television by people who did not know them before and often have not heard them right. And the next development, which follows all too soon, is that the mistakes and misconceptions are repeated on radio and television. So we hear of terrorist actions and bombings reported as "a cynical exercise *on behalf of*" (this or that terrorist organisation), when the speaker meant that it was a crime carried out by that organisation.

The teaching of English grammar seems to have gone to pieces in our schools. This tendency may have started as an extreme reaction against the excesses committed in grammar teaching in our schools in Victorian times and continued well into the early twentieth century. There was in those days a distinct attempt to force upon English some rules that belong to Latin grammar and not to our own. It was a French professor of English at University College, London, who pointed out to me and to all his classes some of the special characteristics of our language, for instance, in the use of prepositions to make complex verbs. One of the more amusing cases of this is our useful verb "to put up with"(meaning to tolerate). Churchill used this one to make fun of the Victorian schoolmasters who further imposed a rule of their own making, that one must never end a sentence with a preposition. Deliberately using for once their preferred order of words in order to ridicule it, he described this rule as "something up with which I will not put". This illustrates perfectly how English, properly used, has a structure, and uses words, very different from Latin and much closer to the Scandinavian languages.

Sometimes changes from older usages may seem legitimate when they achieve some worthwhile aim. Thus, it may be argued that the apparent dropping in recent years by many people of the use of the passive forms of verbs has brought a useful gain of flexibility in speech and writing. And it usually produces shorter phrases. It seems that every verb can now be treated as transitive or intransitive at will. But in these times, when changes are much more readily accepted than of old, many useful distinctions are in danger of being lost by the disregard of old, long-established rules. The rot was obviously beginning to take a hold already in my schooldays around 1920 with all the little boys and girls who promptly said "It wasn't me, Miss" when the teacher demanded to know who it was that caused some noise or commotion. But, at that time, many such mistakes in speech would most likely be corrected at once in class. Nowadays, however, the use of "I" and "me", "he" and "him", "she" and "her" is so commonly confused that even many teachers who grew up in times when the rules were seldom enforced may not easily recognise the proper usage. It may therefore be impossible now to recapture the rule. The translators of the Bible in a slightly earlier generation had it right when they wrote that the prophet said "Here am I, Lord: send me".

Some other disorderly features that have since multiplied have also been around for a long time. The dropping of H's at the beginning of words by speakers (but not in the written language) from most parts of England, though not in Scotland, Ireland or North America, has a long history. Already in the early years of the twentieth century there were R's intruding between words, and within words, where no R should be. And that seems to be becoming ever more frequent. The nation's political and business leaders and even radio announcers and news-readers are often guilty of it. And we have, at least one, weather forecaster who uses his radio appearances to warn the elderly to wrap up because the weather is "roar and cold", and one lady broadcast personality with a name something like Anna Owen always announces herself as "Anna Rowen". We also have many people in England who favour "withdroring" from this or that, particularly from anything to do with "yorup", as they like to call continental Europe and the European Community. This sort of talk has become surprisingly general in England, not least in the House of Commons, but is rarely heard in Scotland, Ireland or America. The "oo" sound seems to cause almost as much difficulty to the inhabitants of south-east, and parts of northern, England as the Scottish "ch". The difficulty seems particularly marked when the "oo" comes before an R: "Your" had long ago become "yor" in the capital.

Jespersen put down most of the blame for the ever-increasing disregard of rules of speech to the universal laziness of mankind, always preferring the easy way. But he was clear that that has not been the only cause. Among the other causes, the urge for freshness and novelty in ways of speaking must be important and is often a cause of pleasure. But it is in the loss of useful distinctions that regrets properly arise, and probably the usual cause of such losses is laziness about thinking precisely and examining any case from various angles. Among the valuable distinctions now on the threatened list, if not yet lost altogether, is that between "may" and "might", as in a recent broadcast reporting a sad failure to secure the release of hostages; the announcer told of some initiative which "may have secured their release". Unfortunately, he had to admit that it had not got them released, although it had been reasonable to hope that the initiative might have worked.

The distinction between "shall" and "will" is now beyond many people's comprehension, because they have been taught to regard these words as no more than alternative ways of forming the future tense. That is a view that misses the point that "will" implies that the human will is involved and "shall" is essentially an order, a command. It is sad to see that even the translators who produced the New English Bible have missed this point. "You shall not kill" is stronger than a statement by me that "I will not kill", however good my reputation for doing as I say. Another failure of this latest supposed modernising of the language of the scriptures is seen in the rendering of Christ's promise to his disciples: "Lo, I am with you until the end of the world", which has been converted to: "I will be with you till the end of the age" - surely a devastating devaluing of Christ's words at a time like the present, when we are surrounded by signs that the age we are used to is ending now if indeed it has not already been succeeded by a new, and more insecure, age.

Meanings are, of course, liable to be completely obscured by the misuse of words. One example of the danger of this arises when the word "*defuse*" (with a long E), describing efforts to take the excitement out of a threatened conflict situation, is sometimes mispronounced in broadcast announcements, as "*diffuse*" (with a short I) meaning to spread through a space, the concentration weakening as the substance spreads. Other, more or less, analogous muddles abound, some of them regrettably exposing their author's lack of command of the language. One case, encountered increasingly in this writer's experience, is the appearance of "*Forward*" as the heading to the "*Foreword*" to a book. And "*Forego*"(= precede) appears not uncommonly in all sorts of texts where "*Forgo*"(= do without) was meant.

Some roundabout expressions arise, no doubt, unavoidably from a speaker's need to give himself time to think, when the next words or phrase intended slip his or her memory. Such popular dodges and tricks have to be accepted, even if they sometimes cause merriment at the speaker's expense. They have given rise to such portentous expressions as "At this moment in time" and "on a daily basis" instead of the simpler "daily".

Other recently adopted fashions may have been introduced in deliberate, kindly intentioned efforts to make the language easier for immigrants to understand and master. This may be the reason for the virtual abandoning of the comparative and superlative moods by some speakers and writers and in announcements. It could even be the supposed justification for such recently heard expressions as "*more small*" (to mean "*smaller*").[10] But

---

[10] BBC Radio 4, 28 February 1994

that is surely going too far. Are we next to have "*more few*"?! In any case, it is noticeable that these things, good and bad, all too easily and quickly become a set fashion that sweeps all before it in these days of mass communication through the media. Examples of these extraordinary trends include some sudden developments of new, distorted and exaggerated pronunciations of common words. Thus, from about late 1993 it quite suddenly became almost obligatory - if one is to be "in the swim" - to pronounce "Primarily" as "*prime-airily*", "voluntarily" as "volunt-airily" (or "volunt-errily"), and so on.

And one, surely very odd, habit seems to be growing in England of putting heavy emphasis, the heaviest emphasis, on the least important words and syllables in any statement. There were some examples of this a long time earlier. We have long been told that everything in the way of fresh air and exercise to green vegetables and fruit are "good *for* you" (whereas surely the emphasis should be on "good"). In the same way, a BBC sports commentator told us that the 1993-4 season had been a "terrible season *for* (a certain Liverpool footballer)".

Another strange characteristic which was already firmly established in my schooldays, and maybe long before that, was (and is) the refusal of even educated folk in England - possibly *especially* of educated people who consider themselves too refined - to so much as try to pronounce the unfamiliar sounds of words from other languages and dialects. One frequent stumbling block of this kind is the Scottish "ch" sound (as in "loch").Many southern English speakers actually prefer to make themselves ridiculous to northern ears by making no distinction between "loch" and "lock". When I was only seven years old, one summer afternoon in a writing class, I accidentally spoilt the sheet on which I was laboriously forming letters and involuntarily expressed my dismay and self-disgust by exclaiming "Ugh" (with the same sound as the "ch" in "loch"). The teacher at once demanded sternly to know "Who made that ugly sound?" and told me "Don't let me ever hear you say that again". That example shows how firmly rooted the prejudice was (and still is). It very likely explains why English people are generally so bad at pronouncing foreign languages. They have been intimidated from their tenderest years about attempting to reproduce foreign sounds both by their teachers and by the jeering ridicule of their equally timid fellows.

There are great pressures to conform probably in all countries, but certainly in present-day England. It is simply not done to dislike cricket, and the schoolboy who doesn't know the names of the latest football heroes is in for some trouble. Nevertheless, and possibly in some way even *because* of this, we do not like being "hide-bound" by rules. Although there are, or used to be, rules in the English language, there are more exceptions in this language than in most. In some cases the exceptions are so many that it is hard to define any rule. Maybe some - possibly many - British people are notably inventive just because of our multifarious, endlessly adaptable language. But some rules are helpful, even though most of them are much disregarded.

One helpful rule about pronunciation, which seems to be by now almost completely forgotten - so far forgotten that many who went to school any time after the nineteen-twenties or thirties may never have heard it - is that a vowel is normally long if followed by only one consonant and short if followed by a double consonant. (Groups of two or more consonants, such as "ch", "ck", "ph", "sh", "th", do not have the same effect as doubles.) Confusion over this is greater, I think, in America than in Britain, although I have met a languages scholar of the nineteen-thirties who had never been taught it. Consequently, it is now about equally common to hear words like "centenary" and "hedonism" pronounced with a short as with the traditional long "E". This creates difficulty for anyone who is not really familiar with the word in question or with its meaning.

The abandoning of rules in English continues, and even seems to gather pace, with less and less regard for ensuring understanding and the awareness of possible ambiguities. One example, which shows several dangerous tendencies, has to do with the words 'innovate' and innovative'. There is no doubt or difficulty in the case of the verb 'to innovate', which is characteristically spoken with most emphasis on the first syllable, although some stressing also of the 'a' in the third syllable is a useful aid to clarity. There is, however, a cult - and possibly a growing cult, popular among the people most inclined to innovate, and who like to be called innovative - of placing all the emphasis on the first syllables of the word so that the rest of the word is spouted out so quickly as to be very indistinctly spoken and hard for the hearer to catch clearly. One may suspect that there may be an intention to convey the impression of a superior community that is always innovating, but the most obvious result is to obscure the origin and meaning of the word even to those who succeed in hearing what was said correctly. Some may think this a reward that is as fitting as it is automatic for what was an attempted pomposity!

Another recent development seems even more extraordinary. This is the pronouncing of the long established word 'precedent' in identically the same way as 'president'. This may be the first recorded pronunciation of the letter 'c' in English as if it were a 'z' - surely an extreme case of disconnecting spelling from pronunciation! It also abandons the long 'e' of 'precedent' in circumstances that might have been designed to confuse.

Epilogue - Language never stands still

There are lots of people, who are always ready to assume that everyone else's choices, and everyone else's pleasures are the same as their own. The gift of imagination is not all that widely distributed, it seems! So we have had an advertisement broadcast on a BBC radio food advice programme which began: "Now, you worry about cholesterol, don't you?" Well. After years of warnings and attacks on obesity, no doubt many formerly carefree people - possibly even most people - have been put into a state of worry about this. But not everybody has the same problem. And not everyone needs to worry about this. Of course, many other examples from other fields could be mentioned. The assumption that we are all the same in our likes and dislikes, and in our worries, becomes very irksome. It cannot be good for society or for our sense of being a community, albeit a richly varied community. And it obviously tends to make some people feel themselves to be "outsiders" and turns some into "loners".

One correspondent, who wrote in to the media in recent years, severely criticised the objective teaching of history in British schools. (Is untrue "history" to be preferred?) He or she wrote: "Are our children not to know that it was we who gave democracy to the world?" Seemingly, the pioneering efforts in ancient Greece and the open air parliament, the Althing, that met yearly in Iceland from AD 960, do not deserve a mention!

Now that we are drawn willy-nilly together with other neighbouring nations in the European community, or are at least influenced by it, it seems that we have to learn what "subsidiarity" means, even though the word was unknown to most people - and even to most English dictionaries - before the politicians introduced it in the late nineteen-eighties and early 'nineties. They gave it a specialised meaning to define the responsibilities of various governing bodies within the European community. It is safe to say that few people in Britain - and very likely few in many of the countries concerned - understood the word before the Danes called it *"naerhed"*, the ordinary word for *"nearness"*: in other words, the principle that governing decisions and regulations should be arrived at as near as possible to the people affected. But it soon became clear that the Major government in England was going to interpret the word just in the way it wanted: i.e. as meaning that all decisions - so far as possible, about everything - were to be taken by themselves, i.e. by the London (Westminster) parliament, and correspondingly by the national parliaments of the other member countries. It should not matter, according to these politicians, what the Scots, with their long tradition and known demand for a Scottish parliament in Edinburgh, might wish (nor, for instance, what risks Scottish children might be exposed to on the roads when going to school in the dark at 9 a.m. because of a ruling of the London parliament that clocks should be kept all the year round on British Summer Time - *alias* Central European Time - an hour ahead of Greenwich Mean Time.) And, on the other hand, it is surely obvious that decisions which affect everyone throughout the length and breadth of Europe should be made by some over-arching body that should certainly be democratically elected. And that surely means a democratically elected parliament for all Europe. Decisions which are only relevant to one nation should properly be made by the national parliament of that country. And decisions about more local matters should properly be made by whatever representative councils are nearest to the people affected - be it a Scottish parliament, a Welsh assembly, or a regional or county council,- and even more local affairs to be ruled upon by whatever appropriate councils exist to deal with them.

INDEX

Accidents, 232-3
Accidents on the hills,
Air Party, privations on the whaling ship, 115, 117, 118, 121, 126-31
Airport security (at Shannon), 88, 101
Air traffic control, 86-7, 96
Arctic hares, 58
Alps, 169, 201, 243-4
American journey, 96-8
Amsterdam, 55
Ancestry, 1-5, 14, 15, 83
Anglo-Catholic influences, 6, 7, 13, 67-8
Anglo-Irish Ascendancy, 104
Anglo-Irish relations, 150-1, 101-2
Antarctic, 112, 124
Archaeology, 187, 225
Archives, 179. 208
Arctic sea ice, 210
Arendal, 236
Argentina, invasion of the Falkland Islands, 216
Arrhenius, Sv., 217
Aspen, Colorado, 225
Astronomical basis of climatic changes, 201
Atlantic air route, Preparations, 54
Atlantic Ocean, Early weather maps, 163
"Auntie Fux", 7
Aviation forecasting (long-distance routes), 85

Aviemore, 231

Bad Eilsen, 159, 162-3, 165, 166-8
Baker, P.E., 189
Bakke, Asbjørn, 240
Balaena, Whaling Factory Ship, 114, 183, 241
Bats, 167
Baur, Prof. F., 154-6
Ben Macdhui, 62
Ben Nevis, 110
Bennett, Donald, 86
Bergen School of Meteorology, 159
Bergeron, Prof. Tor, 159
Berlin, 55
Bessie Foot, 2, 4
Bible, 217
Bjerkenes, Prof. Vilhelm, 53, 159
Black Death, The, 48
Black Forest (Schwarzwald), 244
Blizzard, 229
"Blocking" of the westerlies, 152-3, 155
Bloomsbury set, 5
Boating accident, 93-4
Boating exploits, 90, 93-4
Bombing raids, 10
Book writing, invitation, 195-6
Botanists, 186-7, 200
Bridge, 35
Britton, Captain Graham, 115
Broad Scots, 64
Brutalizing tendency, 9
Bryson, Prof. Reid, 203

Bückeburg, 160
Budget, living on a tight, 38
Budyko, Prof. M.I., 219

Cairngorms, 44, 233
Caithness, 66
"Calendar-bound" seasonal episodes, 153, 156
Calf Love, 67-8
Callendar, G.S., 217
Camp in the snow, 62-3
Canary Islands, 120, 124, 132
Candlemas Day (2 February), 154
Canoeing accident, 90
Capercaillies, 63
Car, first, 62
Carbon Dioxide, 217-8
Cardozo, B.L., 143
Carpentry, 15
Caspian Sea, 211
Catchers (whale), 126
Changes of Climate, causation, 201-3
Cheesery, 34
Christensen, Harald, 240-1
Christmas on the whaling grounds, 123
Circumpolar vortex, 224
Clayton, Dr. Keith, 199
Climate conferences, 201, 204, 210-1, 219-26, 231, 233, 245
Climate inquiries, 178
Climate and history-writing, 200
Climate engineering scheme (Russian Rivers), 211
Climatic Research Unit, 203-4
Climatic scales, 197

Climatic trend, erratic behaviour, 147-9, 203
Climatologist by accident, 178
"Cloudless summer", 137, 139
Clova, Glen, 63, 65
Cod fishery, 147-9, 209
Codes, meteorlogical in wartime, 240-1
Co-education, 11
Colder climate episodes, 209-10, 214
Cold front, 98
Connolly, Dennis, 32-37
Conollys, 37
Conscientious objection, 81-2
Conservation, 210, 217
Consular service examination preparing for, 40
Co-operation with the (whaling) ships' command, 126
Copenhagen, 38
Country life in the west of Ireland, 87-8, 90-3, 100-1, 105
Craddock, James, 158
Cricket, 15
Crisis over funding of Climatic Research Unit, 203-4
Crisis over information for forecasting weather, 119, 121-2, 123
Crossword puzzles, 43
Croydon airport, 53-4
Currie, Bob, 123
Customs examination, 109
Cycling across Ireland, 100
Danish Meteorological Institute, 207, 214

Dansgaard, Prof. W., 188
Day schools, boarding schools etc., 14
Degree crisis, 45
Denmark, 94,163, 235
Derkow, 251
Deutsche Seewarte, 163
Dickson, R.R., 148-9
Doldrums belt, 121
Doporto, Dr. Mariano, 94-5
Dorothy Lamb, 5
Dreyfus affair, 41-2
Droughts in Africa, 212-3
Drought on Ireland's west coast, 93
Drozdov, O.A., 211
Dublin, 4, 17
Dunfermline, 81
Dust in the atmosphere see also volcanoes, 197
Dutch Experience, 158

Eddy, Dr. Jack, 193
Education, Icelandic view, 109-10
Elliotts, 44, 139
"El Niño", 213
Emergency Landing, 126-7
Environment, despoliation of, 249
Eritrea, 212-3
Esbjerg, 239
Ethiopia, 212
Exchange rates, 237
Excitement of the chase (whaling), 128
Faeroes, 147, 210
Fairbridge, Prof. R.W., 220

Falkland Islands Dependencies, 123, 189, 216
Family origins, 1, 4
Family trees, 4
Fata morgana, 75
Fatal accidents, 126
Father, 1
Father's nervous breakdown, 52
Father's disappointments in me, 15
Fevik, 226, 236-7, 239-40, 243
Fines for talking "shop" (bonuses), 123
Fjaerland, 71
Flensing operations, 121-2, 125-6
Flight emergencies, 86, 126-7
Flight forecasts, mass production method, 86
Flights, operational strategy, 135-6
Flint, Prof. R.F., 225
Flohn, Prof. Hermann, 214
Floods, North Sea, 215
Flux, Sir Alfred and Lady, 7-8
Flying boats, 86
Föhn (foehn), 67
Folk weather lore, 153-54
Ford Foundation, 211
Forecasting research at Dunstable, 150, 152,158
Foynes, Co. Limerick, 84-8, 90, 93-6, 98-102, 104-8
Frost, 11, 21
Frydendahl, Knud, 207, 214
Funding of scientific research, 199, 203-4, 206, 210-11

Galdhoppigen, 237

Geographical discoveries in the Far South, 138
Geographical features, deduced from winds, 138-139
Geography, university course, 31
Geologische Vereindgung, 226
German encounter, 8
German language, 162-168
German meteorological society, 165
German navy, 214
Germany, Stresemann government, 17,22
Germany, transference to, 249
Gandhi, 25
Ghent, 234
Girs, 150
Gjessing, Prof. Just, 224
Glaciers, 200-1, 215, 225, 227, 230, 244
Godwin, Prof. Harry, 186
Golf, 61
Gothenburg, 246, 248
Golthard pass, 169
Göttingen, 230
Gränna, 237
Grandfather Lamb (Horace Lamb), 1, 5, 21
Grandmother (Bessie Foot), 2, 4
Gray, 60-2, 64-5
Greenhouse effect, 191, 217, 219
Greenland, 193, 219, 226
Grierson, John, 118, 121, 127
Grimstad, 239
Gudvangen, 237
Guildford, 228-30
Gulf Stream, 147

Gulf Stream switch, 147, 219

Haars, investigation, 57, 58, 59
Hampstead Garden Suburb, 8, 11, 14
Happier times with my father, 16
Hardanger, 33
Hares, Arctic, 58
Harz mountains, 168, 224
Haukelisaeter, 33
Hay fever, 19, 20
Heathrow (London Airport), 110, 112
Heirloom, embarrassing, 185
Helen Palmer (née Lamb), 3, 30
Helga Ramlet, 236-7
Hereford, 6, 238, 241-3
Henry Lamb, 4-5, 8, 16, 83
Herring drifters, a night with, 63
Highland clearances, reminders, 66
Highlands, life in, through recent centuries, 144
Hillside (Angus), 58
Hill walking, 3
Historical records of climate, 204-8
Hitch-hiking, 44-45, 48
Hitlerites, 9
Homes for the elderly, 195
Horace Lamb, 1-4, 22
Hornelen, 47
Horse carriages for tourists, 35
Horse-drawn traffic, 11, 14
House hunting, 180
Huddleston, Trevor, 6, 13
Human influencce on climate, 217-8

Humboldt Current, 213
Hunt, Michael, 168
Hurn Airport, 110, 112
Husfliden, 234
Hypermarkets, out of town, 239

Ibsen, Henrik, 239
Ice Age, 228
Ice recession, 70-71
Iceland
 - family names, 76
 - local effects on winds and weather, 77
 - marriage customs, 76
 - recall from leave, 73
 - swimmng pools, 74
Igloo, 228
Iliffe, Lord, 68-69
Information for forecasting service, 85, 114-5, 119-20
Interglacial warmth, 201
International Association for Quaternary Research (INQUA), 222-8
International Geophysical Year (IGY), 139
International Union for Geodesy and Geophysics (IUGG), 163
International Whaling Commission, 117, 126
Interstadial (Allerød), 228
Intertropical front, 131-2
Invasion fears in Ireland, 91-2
Invasion of Norway and Denmark, 94
Iona Community, 141
Ireland, 83

Irish cattle fairs, 92. 100-1
Irish cross-roads dancing (Kerry), 92
Irish Potato Famine, 102
"Iron Curtain", 165, 222, 230

Jacksons, 176
Johnson, Sir N.K., 81
Jostedalsbre (glacier crossing), 70

Karl Ramlet, 236
Kellocks, 50, 116
Kerry dancing, 92
Kington, John, 207
Kristiansand, 240

Labour Party Conference, 111-2
Lairig Ghru (pass), 44
Lake-bed deposits (varves etc.), 200
Landscape restoration, (Obernkirchen), 167
Languages, 2, 12, 42, 46, 64, 95, 162, 167
Languages - useages in weather forecasts, 253
Lapland, 230
Lettice Lamb, 5, 8
Leggett, D.M.A. (Peter), 26-30
Libya, 173
Liberal Party, Liberal Democrats, 50, 86, 112
"Little Ice Age", 208, 215, 219, 226
Lobster, 61

268

Long-range weather forecasts based on physical arguments, 156-8
Long spells of weather, 153
Lutefisk, 124

MacGillycuddy's Reeks, 90
Maihaugen (open air museum, Lillehammer), 50
Malta, 169
- fine-scale weather development, 172
- unusual summer weather, 171-2
- wild flowers, 170-1
Maltese, unreconciled to their climate, 170
Manchester, 1-3, 7, 12, 14-17
Manley, Prof. Gordon, 163, 198
Mapping past climates, 206-10
Marriage, 144
Mason, Sir B.J., 198
Mathematics, nature of my limitation, 21
Maunder and Spörer sunspot minima, 207-10
Maury, Admiral, 227
M'dina, 170
Meaden, Dr. G.T., 209
Medieval period, a difficult case and diagnosis, 207
Medical check for the Antarctic, 115-16
Meinardus, Prof. W., 114, 135
Mensching, Pastor and family, 166
Meteorological forecasts, 127-8

Meteorological instruments, 117 119, 122
Meteorological Office, appointed to, (43), 75, 108
Meteorological Office, record of sea temperature, 210
Meteorological Office and RAF expansion, 57
Meteorological training in Ireland, 85
Meteorologist with the expedition, 118-20
Meteorology, lessons from *Balaena*, 131-5
Methuens, invitation, 195
Metric system and older units, 179-80, 197
Milk bus (*melkebilen*), 34
Milligans, 140, 144-5
Minister's garage, 62
Mirages, 75
Mitchell, J. Murray, 220, 222
Moir, 145
Moira Milligan, 25, 140-5, 149-50
Montrose, 57
Mother, 5, 14
Mother's family, 5-6
Motor drive (scared), 183-4
Motorized, becoming, 62
Mountain climbers, 3, 62-3
Müllers, the, 50-2
Music, 15-16
Mysak, Prof. L., 148, 219

Namias, Jerome, 157, 225
Nannie, Stevens, 14
Nansen, Fridtjof, 14-15

Napier Shaw Prize Essay
  Competition, 181-2
Napoleon cakes, 37
Nationalistic interpretations, 118
Natural seasons, 153-4
Naval Weather Service, 115, 122-2,
  135-6
Nazi infiltration suspected, 91
Needham, Dr. Joseph, 221
Newfoundland, 96-7
Newnham College, Cambridge,
  5, 195
New York, 203, 207, 220-2
Niemöller, Pastor, 25
Norfolk, 236, 238, 240
Norman, Geoffery and Ruby,
  242
Northern Lights (*Aurora borealis*),
  67
"a little Norway" beyond the
  Highlands (Caithness), 66
Norway, 78-9, 94, 131, 142,
  163-4, 235-41
Norwegian folk tales, 16, 42-3, 79
Norwegian language, 35, 112, 240
Norwegian Youth Hostels
  Association, 46
Norwich, 195, 203
Nuffield Foundation, 199
Numerical weather forecasting, 158

Obernkirchen-choir, landscape
  restoration, 167
Ocean, convergence systems, 133
Ocean currents and their vagaries,
  147-9
Ocean temperatures and world
  weather, 158, 226-7
Olaf Richardson, 6
Ole Ramlet, 236
Opalescence (blue) in the distant
  view in Iceland, 74
Oulu (Uleåborg), 230-1

Pacific Ocean, 212-3
Paisley, 7
"Palatines", 107
Parentage, 1, 5
Paris, Christmas in, 40
Paris Commune, 3
Parkin, Dr. D.W., 197
Parliament, world's oldest, 78
Passport difficulties, 36, 101
Patons, 25, 44, 82
Pearl Harbour, 105
Pedersen, Captain, 118, 121, 123,
  125, 126, 129-31
Peggy Lamb, 4
Persistent musical choices, 16
Peru, 213
Peters, S.P., 53, 85
Petersen Sv., 69
Pfister, Prof. Christian, 207
Piano lessons, 15
Pilot balloon work (complication
  with the planet Venus), 91
Pilot (ship's) embarrassed, 47
Pines and prejudice, 238
Plass, G.N., 218
Pneumonia, 181
Poland, 222-3
Polar explorers, 14, 31

Polar meteorology, working group on, 216
Polio-myelitis, 137-8
Political Life, 250
Politics, interest in, 24, 250
Pollak, Prof. Conrad, 84, 94-5
Pollen analysis, 187, 200, 238
Pollution and stench, 124
Population, 207, 213-4, 217, 223
Postglacial climate development course of, 227
Postglacial warmest time, 227
Presbyterian influence, 24
Priests in Ireland, 83, 105
Promotion interviews, 197-8
Psychology, 13
Ptarmigan, 63

Quaker influences, 6-7, 13, 22, 24, 101-2, 166
Quantock Hills, 13
Quaternary climate discussions, 186-7, 228-9, 232

Racial encounter, A., 111
Radiocarbon dating, 186-7
Rail journeys, 96, 110, 116
Rainfall changes, 207, 212-3
    (as a test of change of prevailing wind direction 207)
Ramlets, 226, 236-7, 240
Refereeing of scientific papers for publication, 185-6, 189
Refugees, 3, 81, 84, 94-5
Reliability of past records, 204-5
Rennell, 227
Resignation crises, 81-2, 108-9

"Return period" (average), 178
Reunions with the grannies, 175
Revanchism in the inter-war years, 22
Reynolds's, 242
Richardson, L.F., 6-7
Ringebu, farms and stave church, 49
"Roaring Forties and Fifties", 120, 123
Rockefeller Foundation, 204, 206, 208, 315
Rocky Mountains, 225-6
Rodewald, M., 163
Rome symposium, WMO/UNESCO, 225
Rose Brooks, 17
Rothiemurchus, 44
Royal Meteorological Society, 200
Royal Swedish Geographical Society's *Vega* medal, 182
Russia, 222-3
Russian emigrés, 12
Russian rivers scheme, 211-2
Rutting stags, 63

Saas, 29-30, 244
Saeter, night on, 48-9
Safety of aircraft, 88, 98-9, 107, 126-7, 129-30, 135-6
Sahel, 213
St. Swithin's Day (15 July), 153-4
Sandbjerg, 234
Sapper, Karl, 188
Sauna, 231
Scandinavian influences, 7, 25, 26, 44-5

Schaumburg-Lippe, 160
Scherhag, Dr. R., 158-9
Schneider, Dr. S., 190-1
Schooling, critical points, 19-22, 23, 194-5
Schove, D.J., 221
Schweizer, Dr. A., 25, 166
Scirocco,
Scotland, 57-72, 238, 243
Scots, Broad, 64-5
Scott, Captain R.F., 14
Scott, Peter, 14
Scottish Country Dancing, 65
Scottish weather - an easy way for a meteorologist to gain a reputation for reliable forecasting, 66-7
Scouting, 20
Sea breezes, 58-9, 91, 172-3
Sea floods, 215-20
Sea ice, 147-9, 210, 215
Sea level, changes, 214-5, 227
Seasickness, 63, 69, 124
Seasons, natural, 153
Sea temperatures, 210, 212-3, 218, 227, 240
Seattle, 226
Second home, experience of, 242-3
Setesdal, 237
Sex and modesty, 37, 70
Shackleton, 14
Shell, 199
Shelter Stone, the, 233
Shetland, 149, 210
Shooting expedition, trial, 58
Shotton, Prof. F.W., 225

Siberian rivers, scheme ("climate engineering"), 211-2
Silesia, 223
Siljan, lake, 237
Simonstown, 115, 122, 123, 135
Simpson, Sir George, 43, 138
Singularities, 153
Skagerrak, 237
Skating, 21, 87
Sleep, tactless, 111
Sleeplessness, 89
Smugglers, 109
Snuff, 17
Social Democracy, 251
Social divisions, 250
Societas meteorologica palatina, 152
Society of Friends (Quakers), 7, 14, 166
Solar influence, 192-3, 210
Somalia, 213
Southern Ocean, 134-6, 143
Spanish Armada storms, 208-9, 215
"Spare the rod and spoil the child", 12
Speyside, 44, 146
Spörer minimum of solar disturbance, 206, 210
Squirrels, 238
Stevens, Florence, 14
Stockport, 1-3
Stockholm, 182
Storminess, 214-5
Subtropical convergence, 133
Sudeten German refugee, 81
Sultan of Zanzibar, 8

Sunburn, 131
Sunspots, 191, 192-3
Surströmming, 245
Sutcliffe, Prof. R.C., 151, 155, 158, 189
Sutton, Sir Graham, 195-6, 199
Sweden, 32, 36-7, 46, 209, 230, 251
Swell, Atlantic, 91
Switzerland, 28, 244

Table manners, 19
Talking "shop", 123
Telephoning in Ireland (1940s), 99
Temperature, world, trend of, 218
Tenerife, 120
Tent trouble, 164
Thames Barrier, 214
Theoreticians and field scientists in climatology, 200-1, 225
Thorarinsson, Sigurður, 189
Thrashings, 12, 19
Time-keeping in wartime Ireland, 109
Tønsberg, 116
Tourist literature, 171
Trade secrets, 129, 133-4
Train travel, wartime Ireland, 100
Training meteorologists, 53-4
"Tramp ship" crossing, 46-8
Trans-Atlantic "Clipper", first encounter, 84-5
Trans-Atlantic "Clipper" crossings, 96-8
Tree rings, 249
Trinidad, 160-1

Trinity College, Cambridge, 2, 12, 22
Trondheim, 48
Trouton, Mr. (Chairman, United Whalers Ltd.), 117, 118, 121, 123 126, 127, 128
Tycho Brahe, 209

Uncle Reggie (Reginald Brierley), 5, 241
Undset, Sigrid, 25
UNESCO, 225
United Nations Environment Programme, 211
Units (old scientific units of measurement), 197
Unruly house, 45

Valdres, 241
Valentia, 93
Venice, alarm, 174
"Vetting" research papers for publication, 185
Victoria, Lake, variations of, 212
Volcanoes, volcanic dust eruptions: impact on climate, 188-191

Wager, Prof. L.R., 189
Wales, 3, 6, 242
Walter Lamb, 5, 12, 16, 17
Warren, Grace and Sam, 242
Warming, global, 191, 217-9
Warming of world's oceans and energy of atmospheric circulation, 214
Warm epoch, medieval, 207-8, 215

Warmth of the twentieth century, 218
War psychology and its aftermath, 22
Wartime hatreds, kept alive, 22, 40-2, 55-6
War winters, severe, 87, 88
Water vapour, 215
Weather analysis on board the whaling factory, 119-20, 121, 126-7
Weather messages, 119, 121
Weather (and wind) pattern classifications, 151-3
Weickmann, Prof. L., 165
Westerly winds, varying prevalence, 155
Wexler, Prof. H., 188
Wh/F *Willem Barendsz*, 135, 143
Whaling mystique, 117, 126, 130
Whaling station (Icelandic), 74
White heather, 65
Wild cat, 65
Willett, Prof. H.C., 193
Williams of Hillside, 65
Wind circulation, global, 212, 224
Wind systems, northern and southern hemispheres compared, 182
Winds at 3,000 metres, 95-6
Winter sports, 21, 78-9, 87, 173, 228
Wolfson Foundation, 203
World Meteorological Organization (WMO), 210-11, 216, 225

Year layers, 200

Yorston, George, 65

Zuckerman, Lord, 214